Stuxnet:
The Revenge of Malware

How the Discovery of Malware

from the Stuxnet Family Led to

the U.S. Government Ban

of Kaspersky Lab

Anti-Virus Software

By Roman Poroshyn

Contents

Author's Note

On June 4, 2019, Kaspersky Lab announced its rebranding. The new name and the new logo had been revealed to the public. According to the company, it has become a global company. On that stage, Kaspersky Lab has outgrown the old anti-virus lab, so the word "lab" disappeared from the name. Now it is simply Kaspersky, and it is not written using Greek alphabet letters anymore. The renewed company, under the name Kaspersky, promises its four hundred million users and two hundred seventy thousand corporate clients worldwide not simply "cybersecurity" but complete "cyber-immunity".

Most of the events described in this book took place before 2019. That is why the old name of the company, Kaspersky Lab, has been used, where it was appropriate.

Introduction

Duqu, the infamous malware from the Stuxnet family, made its return. First discovered in October 2011, Duqu was actively spying on digital certificate issuing authority in Hungary. As soon as it was exposed, the malware was abandoned by its originators. Information security companies added it to a "blacklist" database of anti-virus software. After that initial discovery, Duqu had been out of the spotlight for almost four years, until in 2015 of the biggest names in cyber-security, Kaspersky Lab, announced that it had detected the presence of the Duqu malware.

Duqu was always after the high-profile targets but this time the malware was definitely in the wrong place. Duqu had been found alive and well within Kaspersky Lab's very own corporate computer network. That discovery had big implications not only for the ways anti-virus software detects intrusions but also for Kaspersky Lab itself. Usually malware, after being detected, has no means of fighting back. It was not the case with Duqu. Two years after its exposure, the Duqu malware had its revenge.

In 2017, the USA, UK, and some other European countries issued a ban on the use of Kaspersky Lab computer security software on their government and military computers. Their decision was based on information presented to them by the originators of the Duqu malware, which had been quietly stealing data from inside of Kaspersky Lab's corporate computer network.

In an unbelievable turn of events, detection of malware by one of the most recognizable names in the information security industry led to a removal of Kaspersky Lab's anti-virus software from computers that by association require the most protection because they keep top secrets and are constantly targeted by adversaries.

To better understand how it could have happened, we need to explore the story about the Duqu malware, the only malware that had its revenge.

Chapter 1.
The return of Duqu

On June 10, 2015, Kaspersky Lab issued a press release with an intriguing headline, which began with the words "Duqu is back". The press-release acknowledged that Kaspersky Lab's corporate network was attacked by new malware, loaded with three zero-day vulnerabilities. In the same press-release, Kaspersky Lab emphasized that it was not the single target of the malware. Other targets were computer networks in three luxury hotels in Switzerland. During 2014 and 2015, those hotels hosted foreign diplomats representing countries P5+1 at series of nuclear negotiations with Iran. The abbreviation "P5+1" refers to five permanent members of the United Nations Security Council (France, the United Kingdom, China, Russia, and the United States) plus Germany. The other target of the malware were events in Poland attended by many foreign diplomats in January 2015. It was a commemoration of the 70th anniversary of the liberation of the former German Nazi concentration and extermination camp Auschwitz-Birkenau.

Members of the general public, who were not following Kaspersky Lab news daily, had learned about the return of Duqu from a big splash of articles on the Internet. Even a brief look at headlines shows that the focus of attention had been split between two main news: first - previously unknown malware infiltrated the internal network of the well-known anti-virus company, and second - nuclear negotiations between the Western powers and Iran were "bugged". Here are some examples of the headlines: "Kaspersky finds new nation-state attack – in its own network" (Wired magazine), Stepson of Stuxnet stalked Kaspersky for months, tapped Iran nuke talks" (Ars Technica), "Security firm says spyware targeted Iran nuclear talks" (USA Today).

From interviews given by Costin Raiu, the director of Global Research and Analysis Team for Kaspersky Lab, and Vitaly Kamluk, a principal security researcher at Kaspersky Lab, more details have emerged. The malware infected one of the computers in the Asian-Pacific regional office of Kaspersky Lab. Then it granted itself the rights of the domain administrator, including access to Microsoft Software Installer (MSI). Using the MSI, malware installed itself on other networked computers, slowly but surely making its way from Asia to the Kaspersky Lab's headquarters in Moscow, Russia.

The cyber-attack was discovered in the spring of 2015 when an engineer from Kaspersky Lab began testing a prototype of anti-virus software. This newest software had been designed to identify the presence of the most sophisticated type of malware - advanced persistent threat (APT). During the test, the engineer's very own computer demonstrated some odd behavior. In the beginning, nobody could believe or even suspect that the computer had a virus. That particular computer was fully patched, had all recent updates, and was an integral part of a highly protected corporate network. That was the network of the world-leading anti-virus company after all, where all known viruses had been accounted for, quarantined, and closely monitored in a protected "sandbox environment" with no chance to escape.

After multiple checks, researchers zeroed in on a single module, which disguised itself as a part of legitimate Microsoft software. A closer look at the discovered sample revealed the signs of intrusion by a highly sophisticated malware platform. Judging by the complexity of the design and unorthodox approach, used by its originators, the creation of that type of malware must be sponsored by a nation-state. That was a conclusion of Kaspersky Lab experts.

The total number of infected computers at Kaspersky Lab was never clarified (somewhere around "couple dozen" according to a journalist Kim Zetter). During several weeks, security researchers were able to watch how the malware was traveling through the network, installing inside infected computers just one module or the full-scale 19-megabyte package, slowly siphoning data, collecting it, repacking it inside image files, and sending to its command-and-control servers.

At the same time, Kaspersky Lab engineers continued to look for the initial point of infection. Somehow their activities triggered hackers. Four hours before researchers remotely entered the first infected computer in the Asian-Pacific office, the malware originators wiped out the browsing history and emptied the mailbox on that computer. After that, they abandoned their malware by stopping all communications. All command-and-control servers went dark.

The detailed analyses of the malware by Kaspersky Lab revealed similarities between the newest malware and the infamous member of the Stuxnet family – Duqu. To have its assumptions independently confirmed, Kaspersky Lab shared samples of the malware (now named Duqu 2.0) with its colleagues. The release of their conclusions was coordinated with Kaspersky Lab and appeared on the same day, on June 10, 2015.

Symantec confirmed findings by Kaspersky Lab in its report "Duqu 2.0: Reemergence of an aggressive cyberespionage threat". Also, Symantec identified additional targets such as two telecoms operators, one in Europe and the other in North Africa, a manufacturer of electronic equipment in Southeast Asia, and single computers in the US, Great Britain, Sweden, Hong Kong, and India.

Ars Technica was one of the first to draw a connection between the newest version of Duqu and Israel. To be honest, that was a rather easy thing to do. About three months earlier, in

March 2015, the Wall Street Journal published the article "Israel Spied on Iran Nuclear Talks with U.S." In this article, Adam Entous quoted former and current U.S. officials saying that "Israeli intelligence agencies sweep up communications between U.S. officials and parties targeted by the Israelis, including Iran.".

Both, the nature of targets and sophistication of the newly discovered malware, combined with already known information about a country not involved in nuclear negotiations with Iran but spying on other participants, singled out the one nation-state – Israel. The press-release by Kaspersky Lab did not name the nation-state responsible for the malware but rather clarified which tools of the trade were used by Israel.

Among others, the blog post "Duqu 2.0" by CrySyS went almost unnoticed by the general public, which was not fair. CrySyS or CrySyS Lab are abbreviations for the Laboratory of Cryptography and System Security at the Budapest University of Technology and

Economics in Hungary. In May 2015, Kaspersky Lab shared the samples of the Duqu 2.0 malware with the CrySyS Lab. The choice of the CrySyS Lab for peer review among other easy recognizable names in the information security industry was not accidental. Researchers from the CrySyS Lab were the first who discovered the original Duqu malware back in 2011. Now they had a chance to take a look at the new and improved version of Duqu.

On June 10, 2015, joining others in the coordinated announcement of the discovery of the newest threat, the Laboratory of Cryptography and System Security responded by pointing out the connection between Duqu 2.0, Stuxnet, and Duqu 1.0, with the following statement "... now we have a new member of the same family!"

Let's take another look at the family of Duqu 2.0, in particular, at how the original Duqu 1.0 came to the spotlight.

Chapter 2.
Duqu 1.0 - the Stuxnet's Spy

July 2010 had been considered a victorious time for the information security industry. The Stuxnet malware was discovered, hunted down, and neutralized. The command-and-control servers of the Stuxnet computer worm were shut down. The information security community had a good solid reason to believe that the case of Stuxnet was closed forever. That good feeling of a complete and final victory did not last long. Things changed drastically on October 14, 2011. That day, the Laboratory of Cryptography and System Security (CrySyS) at Budapest University of Technology and

Economics in Hungary shared a sample of previously unknown malware with Symantec. The sample was accompanied by a 60-page report.

Dr. Boldizsár Bencsáth, a.k.a. Boldi, and his colleagues from the CrySyS Lab, who discovered the new computer threat, named it "Duqu" by the prefix – "DQ" that the malware gives to files it creates. The researchers from Symantec were struck by the same thing, which shocked Dr. Bencsáth. The programming code of the Duqu malware had surprising similarities with the programming code of Stuxnet. It led Symantec to the assumption that the new threat was created by the same people who originated the Stuxnet computer worm. As always Symantec had been politically correct by adding the following reservation in parentheses, "The threat was written by the same authors (or those that have access to the Stuxnet source code)."

The even bigger shock was an original conclusion that Duqu was created after the Stuxnet computer worm discovery became public. At that time, Stuxnet's command and control servers were shut down, the United States and Israel were accused of creating Stuxnet to stop the Iranian nuclear program. Everybody thought that Stuxnet's creators ceased all their activities. Liam O Murchu, a supervisor of Symantec Security Response Operations, said in his interview to Wired.com, "We thought these guys would be gone after all the publicity around Stuxnet. That's clearly not the case. They've clearly been operating over the last year. It's quite likely that the information they are gathering is going to be used for a new attack. We were just utterly shocked when we found this."

At the time of its discovery, the first known attack by Duqu was traced back to August 2011. The target was the Hungarian company NetLock. Since the connection between Stuxnet and Duqu had been already established, the interest of Duqu originators in the target's business was easy to understand. NetLock was one of the two Hungarian companies, which had the authority to issue digital certificates. It came as another painful reminder about Stuxnet, which pretended to be legitimate software by using two stolen digital certificates. However, Stuxnet's certificates did not have its origin in Hungary. They belonged to companies located in Taiwan and were issued by VeriSign.

Duqu itself was not a stranger in using a stolen digital certificate. Its originators supplied the malware with an authentic digital certificate stolen from C-Media Electronics Incorporation with headquarters in Taipei, Taiwan. As it was the case with Stuxnet, the stolen certificate was originally issued by VeriSign. From the previous experience, Symantec knew that a computer threat with more than one stolen digital certificate was much more than a remote possibility. Fortunately, it was not the case with Duqu. No other certificate was found at Duqu's disposal.

Symantec, which at that time had recently purchased VeriSign, remembered its lesson with Stuxnet's stolen digital certificates. Back then it took almost a month to revoke the first stolen digital certificate used by the Stuxnet computer worm. This time the compromised certificate was revoked on October 14, 2011, the same day Symantec started to investigate the first sample of Duqu.

Duqu was discovered, and the hunt began. The first major finding related to Duqu was that it did not carry any dangerous payload like Stuxnet. Instead, Duqu acted as a spy that prepares a ground before the main troops enter the battle. Duqu collected intelligence data such as design documents from industrial control system manufacturers. Symantec detected six organizations, which were attacked by Duqu: one organization with locations in France, Netherlands, Switzerland, and Ukraine, another organization in India, two other organizations in Iran, another organization in Sudan, and the other organization in Vietnam. Other information security companies found Duqu's presence in Austria, Hungary, Indonesia, United Kingdom, and some other Iranian based organizations that were not previously discovered by Symantec.

Duqu was spreading around through targeted emails with an attached Microsoft Word document. After the receiver of email opened the document, Duqu used a previously unknown breach in Microsoft Windows defense, a.k.a. "true type font parsing vulnerability", to install itself in a computer. In its Security Advisory, Microsoft admitted that using this vulnerability attackers could create new user accounts with full privileges. Those privileges meant that hackers could install other programs of their choice, view, modify, and delete any data.

After infecting a computer, Duqu established the Internet connection with its command and control server. Through this connection, Duqu would receive instructions on how to download another module, so-called infostealer. This new module would quietly record keystrokes of usernames and passwords and collect other sensitive system information. Duqu spread through a network by using a network file sharing protocol SMB (Server Message Block). The same way malware delivered stolen information from computers within the network, which did not have an Internet connection, to a computer connected to the Internet.

The hunt continued. In October 2011, Indian authorities seized computer equipment from a data center in Mumbai used to host Duqu's command and control server. Still, the originators of the Duqu malware did not give up. Almost immediately, another command and control server became operational in Belgium. Symantec notified Combell Group, the Belgium web-hosting company, about those suspicious activities on November 1, 2011.

Even though Combell Group officially announced the shutdown of the server, for at least two more days the server had been fully operational, and very busy by sending emails to other computers. This is probably how the stolen information was retrieved by Duqu's creators. To avoid future detection, all email addresses and other details of communications were deleted immediately after the completion of data transfer. According to Benjamin Jacobs, a Chief Technology Officer of Combell Group, the company was investigating the situation related to continuous use of the server after it was officially shut down.

A different explanation of the above-described events came from Tom De Bast, Combell Group's business development manager. In his interview with the UK-based Datacenter Dynamics, Tom De Bast said that the final decision to take the server offline was made solely by the management of the company. According to him, the Combell Group never received any requests from authorities. Datacenter Dynamics quoted De Bast as saying, "We heard about the Symantec report on 1 November, then two days later, after not hearing from the server's owner or the police or Belgium Computer investigators which look into these things in Belgium, we shut the server down."

Back then, researchers at Symantec were convinced that Duqu was used to create a groundwork for Stuxnet-like future attacks on industrial infrastructure and manufacturing plants. It made perfect sense because Stuxnet was more than once called a blueprint for future cyber-attacks. With such a huge interest in Stuxnet from the general public, it was expected that after the bad publicity the next cyber-attack would be delayed at least for a while. In real life, there was no break at all. The cyberwar did not stop even for a moment.

As we had learned later, the cyberwar started earlier than everybody thought. At the end of 2011, there was not enough sound evidence of Duqu's presence before Stuxnet's attacks on Iran. That is why the Duqu malware was often called by media the "son of Stuxnet". Even though the compilation time of the programming code of Duqu was later than one of Stuxnet, some experts were not convinced that it was true.

On November 2, 2011, U.S. Air Force cyberspace officers Jeremy Sparks, Robert M. Lee, and Paul Brandau published their vision in SC Magazine. In their post "Duqu: father, son, or unholy ghost of Stuxnet", they wrote that "Duqu has been hailed as the "son of Stuxnet" but it is possible that it is instead the "father of Stuxnet." In their opinion, there could be more undiscovered yet versions of Duqu, including those created before Stuxnet.

Only in March 2012, in its press release, Kaspersky Lab stated that "...the first trace of Duqu-related malware dates back to August 2007. The company's experts have recorded over a dozen incidents involving Duqu, with the vast majority of victims located in Iran." It meant that Duqu was around long enough to prepare the way for the Stuxnet computer worm and continued to pave the way for new cyber-attacks even after Stuxnet's existence was revealed to the world.

Chapter 3.
Duqu 2.0 at Work

As researchers from Kaspersky Lab continued to analyze the Duqu 2.0 malware, they were finding more and more tools of the cyber-spy trade: from a stolen digital certificate to three Windows zero-day vulnerabilities. The digital certificate was issued by Verisign to Foxconn Technology Group based in Taiwan. That stolen digital certificate was the absolute must-have for Duqu 2.0. Since the 64-bit Windows operation system requires every driver to be digitally signed, for this occasion, drivers of Duqu 2.0 would present Foxconn digital certificate as proof of their legitimacy.

As for zero-day exploits, more than likely,

one of them, hiding inside a Microsoft Office document attached to an email, was used during the initial infection of a computer of Kaspersky Lab employee in the Asia-Pacific office. Other zero-day vulnerabilities were needed to grant the malware domain administrator rights and allow it to burrow deep inside the corporate computer network.

This is how the process works. Duqu 2.0 uses Microsoft Windows Installer to migrate from computer to computer pretending to be Windows update. When a package containing a payload arrives into another computer, it is remotely activated by using another Windows service known as Task Scheduler. The technique itself is not new, it was already used by Duqu 1.0 in 2011. Depending on how important the newly infected computer is for the attackers, Duqu 2.0 uses one of two packages, which it has at its disposal. The basic package contains a remote backdoor, which would allow access to the infected computer on an as-needed basis. The main package has a full-scale set of surveillance modules. All of the modules live entirely in the computer's memory. This means no single file would be written on the disk, but it would take 18 megabytes of space in memory.

It appears that originators of Duqu 2.0 truly believe that "the best defense is a good offense". The first thing that Duqu 2.0 does after penetrating the computer, it launches several attacks against its main enemy - anti-virus software. One of the targets is the "avp.exe" process [anti-virus product executable]. While preparing for the attack, Duqu 2.0 maps the location of Kaspersky Lab antivirus. The malware cannot physically move itself to the same location, but it can do a few tricks to pretend that it exists in the same trusted location as anti-virus. Duqu 2.0 applies two patches to "apv.exe", which would be used for future communications.

Using those patches as a cover for its real location, Duqu 2.0 tries to communicate with Kaspersky Lab minifilter driver. The name of the driver is "klif.sys" [Kaspersky Lab Intruder

Filter], which is an advanced virus interceptor. From outside it looks like one of the modules of the trusted anti-virus product, "apv.exe" itself, communicates with the advanced virus interceptor, "klif.sys". This is how the Duqu 2.0 malware pretends to be a part of anti-virus software. After that, the malware can register itself as a trusted process. As the registered trusted process, the malware is excluded from the anti-virus scanning, process monitoring, and firewall restrictions.

Nevertheless, it is not enough for Duqu 2.0, it insists on the highest protection status, which could be provided by Kaspersky Lab software. That protection is essentially a self-defense feature, which is used to protect anti-virus from advanced types of malware. An advanced malware often attempts to kill all security software processes running inside the computer by using unpatched vulnerabilities in the Windows operating system. In the case of Duqu 2.0, the Duqu 2.0 malware cares less about being shut down by another malware. It has something different in mind. After the special protection status is granted, now nobody, even the system administrator, would be able to stop any process initiated by Duqu 2.0.

Kaspersky Lab has stated that mentioned above attack alone would not succeed because since 2010 its anti-virus also checks a digital signature assigned by Kaspersky Lab to its drivers. Duqu 2.0 drivers had no ways of passing this additional check, and shortly would be detected by Kaspersky Lab advanced virus interceptor "klif.sys". In anticipation of this turn of events, Duqu 2.0 originators crafted another sophisticated attack. The main target of that attack by the malware becomes the "klif.sys" itself.

The narrow-pointed attack begins by using Windows kernel mode zero-day vulnerability. The attackers want to load and execute a malicious driver "KMART.dll" directly in kernel mode. Kernel mode is used by a Windows operating system to run the most trusted processes, and it is completely separated from user mode. For those reasons, running malware in kernel mode gives its originators unmatched opportunities to proceed undetected and alter the Windows operating system itself. That is exactly the goal of the "KMART.dll" driver.

Once executed, "KMART.dll" turns its attention directly to Windows API (Application Programming Interface), which is responsible for communications between software applications and operating system. Through API, "KMART.dll" accesses address tables used by certain functions of "klif.sys", such as "Get Current Process ID", "Lookup Process by Process ID", and "Get String Reference". "KMART.dll" modifies addresses to include locations of the Duqu 2.0 malware. From now on, the "klif.sys" driver thinks that all malicious processes initiated by Duqu 2.0 are coming from the same location as the trusted process "avp.exe" or anti-virus itself. And Kaspersky Lab admits, that second attack would succeed. As a result, Duqu

2.0 is excluded from anti-virus monitoring, and the malware can fully utilize all its payloads without the risk of being detected and stopped.

After establishing and securing itself, the Duqu 2.0 malware opens its payload and starts network and domain discovery. It lists all servers (paying special attention to servers with high uptime), records IP addresses, usernames with associated passwords, collects information about operating systems, lists all running processes, copies file names and directories.

All collected information is disguised as Windows file-sharing traffic and send to all-in-memory storage located on a server, which has an Internet connection. There the stolen data is repacked inside image files. The same way it was done by its predecessor – Duqu 1.0. To avoid attracting attention and its safety, the malware does not have any addresses of command-and-control servers hardcoded in its programming code.

Without the hardcoded address of the command-and-control server, Duqu 2.0 is unable to initiate contact with its originators. It waits for them to make a connection. Duqu 2.0 is dormant inside a server with an Internet connection. Everything changes, when the server receives an external communication containing a special keyword. In the case of Kaspersky Lab, the following keywords had been used: "romanian.antihacker" and "ugly.gorilla". When the keyword has been received, Duqu 2.0 awakes. It checks the keyword. If it matches one of the hardcoded keywords, Duqu 2.0 sends the prepared "image" files to the IP address from which the keyword came. Next time the IP address will be different, but the keywords remain the same.

The keywords "romanian.antihacker" and "ugly.gorilla" would accompany updates and orders coming from the malware originators. Upon receiving new orders, Duqu 2.0 would be taking screenshots, copying, overwriting and even deleting specific files of interest or creating new ones. Again, the data would be collected, repacked as image files, and stored until the next connection would be established.

Even though the Duqu 2.0 malware is classified as a part of the malware group collectively known as ATs (Advanced Persistent Threats), Duqu 2.0 stands alone from the bunch. The reason for that is that Duqu 2.0 does not fit the definition of the APTs malware group. Duqu 2.0 does not have a persistence mechanism. It means that in the event of shut down or reboot, an infected computer would disinfect itself.

Creators of the Duqu 2.0 malware have been aware of that scenario and came up with a solution. Designed to attack a network of computers, Duqu 2.0 stores a few selected drivers inside servers with an Internet connection. Those servers traditionally have the longest uptime and more than likely would not reboot simultaneously. This way Duqu 2.0 keeps its continuous presence within the network with an ability to infect again any computer, which was restarted.

This is how Costin Raiu, from Kaspersky

Lab, explained that in his interview with Ars Technica, "The only chance to get rid of the whole thing is just to reboot everything at the same time. So if you simulate an entire power failure for the building, then you will get rid of everything pretty much everywhere. If you reboot them one by one, then the malware will keep jumping from one computer to another and you will achieve nothing." To get rid of Duqu 2.0, a power outage had been simulated for the Kaspersky Lab's corporate network. Then the infected servers had been found, and Duqu drivers had been removed.

Costin Raiu, who is the director of Kaspersky Lab's Global Research & Analysis Team, could not resist but to compliment creators of the Duqu 2.0 malware, "With these guys, they were so confident that they can produce an espionage platform that runs entirely in memory. I believe that this kind of philosophy and thinking is a generation ahead of pretty much everything else that we've seen in the APT world."

Chapter 4.
Anti-virus companies
have been targeted

All is fair in love and war. The famous statement by John Lyly dates back to the XVI century. Things have not changed since then. When it comes to cyber-war in XXI century, everything is acceptable, all is fair. It is especially true if we are looking at big players such as nation-states.

Malware produced or purchased by nation-states comes supplied with stolen digital certificates, fully loaded with zero-day vulnerabilities, and pretends to be legitimate software updates. This type of malware is called the APT (Advanced Persistent Threat). By definition, APT is the most capable type of computer viruses that is hard to detect and difficult to disable. The development of APT requires much more than basic knowledge of programming. It demands expertise, many hours of code writing, and an unlimited budget. That is why there is no surprise that APT comes from arsenals of nation-states.

After spending so much money and resources, nation-states would want something big in return. That big is nothing less than unrestricted ability to successfully target any organization or individual anywhere in the world. It is fair to admit that nation-states had enjoyed this ability for a long time until recently. During the after-Stuxnet years, new and improved anti-virus software became the major obstacle for APTs. To overcome this hurdle, the list of targets for the most sophisticated malware has been extended from adversaries, rivals, and political opponents to include anti-virus companies.

This is how Vitaly Kamluk, a researcher from Kaspersky Lab, explained it in his interview with the Ars Technica in 2015. He said, "We see this battle of arms race emerging and now it involves some kind of confrontation between the security industry and nation-state sponsored spies... There was kind of an unspoken rule not to attack the security industry. But now we see they are stepping on this territory. They are trying to step on our land and they are trying to ruin the last island of safety..."

On June 22, 2015, seven days after the Ars Technica posted the statement from Vitaly Kamluk, the Intercept came up with proof. The article "Popular Security Software came under relentless NSA and GCHQ attacks" was based and accompanied by the documents provided by Edward Snowden, the National Security Agency (NSA) whistleblower. The article sheds light on the variety of ways that top U.S. and British spies had been implementing their attempts to bypass Kaspersky Lab's anti-virus software.

According to the article, the Government Communications Headquarters (GCHQ), the agency in charge of signals intelligence in the United Kingdom, approached the Kaspersky Lab software the same way as any respectful information security company would approach malware. GCHQ worked on reverse engineering of the anti-virus program, trying to recreate it in order to better understand its inner workings. The main objective was not to reinvent the off-the-shelf product but to find vulnerabilities of the Kaspersky Lab software. That is where things became complicated in more than one way, including the legality of the approach chosen by GCHQ.

The practice of reverse engineering is perfectly legal when it is applied to malware. On the other hand, when it comes to copyright-protected products including computer security software, it is illegal in most countries around the world. Great Britain is not an exception.

That is why GCHQ needed special permission to break the law. British spies filed an application with the Secretary of State to obtain a warrant on reverse engineering of security software products. The copy of that application obtained by the Intercept was not an initial application. It was the application to renew the existing warrant, which was about to expire.

The heading of the mentioned above application, dated June 13, 2008, included words such as "Top Secret" and "UK Eyes Only". The application itself requested to extend a warrant for activities such as "modifying commercially available software to enable interception, decryption, and other related tasks, or 'reverse engineering' software." To justify this request, the application also included some bites of information related to how the previous warrant was used. Also, there was some bragging about successful "computer network exploitation (CNE) operations". One of the examples of a computer network exploitation was spying on the Pakistan Internet Exchange, which provides access to the Internet for the majority of users in Pakistan. That secret operation became possible as the result of reverse engineering of the Cisco routers' software used by the Pakistan Internet Exchange by GCHQ.

The National Security Agency (NSA) did not simply sit and watch its British counterparts. It decided to use a different approach in its fight with the Kaspersky Lab software. In 2008, NSA concentrated its efforts on intercepting communications between Kaspersky Lab software and its server. The information exchanged during those communications included a variety of data collected by an anti-virus program about a user's computer. According to the NSA, some of that information could be used to identify individual computers and possibly even their users.

As it was expected, Kaspersky Lab strongly denied any possibilities of the NCA claim. The main argument of the company was the use of strong encoding combined with depersonalization of all information sent by a user's computer to their company server. Despite the sound statement by Kaspersky Lab, the Intercept came up with a series of tests of its own. Based on the results obtained, the Intercept noted that during its tests of Kaspersky software at least part of the information had been sent without any encryption. That information included hardware configuration and the names of software installed on a computer.

Another revelation by the Intercept was the existence of another big-scale spying operation - the Project CAMBERDADA. A PowerPoint presentation on this project from 2010 exposed another tool used by the NSA. Again, it was about intercepting electronic communications but not between computers. This time it was an interception of emails from security experts around the world to the leading anti-virus companies containing information about newly discovered malware. Upon receiving such hints, NSA employees routinely checked if Kaspersky Lab software started to catch those newest viruses. If the answer was negative, then the NSA would consider using that newly discovered malware for its own purposes until it would become detectable by anti-virus software.

The presentation had another slide, which requires special attention. The heading of that slide is "More Targets", which meant that it was an updated list of targets for 2010. That target list included a total of twenty-three security companies based in different countries. The Intercept pointed out that not all of the major companies were listed as targets. For some unknown reason, the list did not include the US-based McAfee and Symantec and UK-based Sophos anti-virus companies.

After reading the Intercept article and accompanying documents, it becomes obvious that Israel was not alone in its interest in Kaspersky Lab software. At the same time, no other malware, except Duqu, was able to go so far as to penetrate all defenses of the most reputable anti-virus company.

The Revenge of Malware

Chapter 5.
Israel Denies but
Media Says Otherwise

In June of 2015, as the news about Israel's involvement in the cyber-attack against Kaspersky Lab kept spreading, the state of Israel denied any possible connections to the Duqu 2.0 malware. During the interview with Israel Army Radio, Tzipi Hotovely, then-Deputy Minister of Foreign Affairs, said, "There is no basis for the international reports claiming Israel was involved in the matter [such as spying on the nuclear negotiations with Iran and the Duqu 2.0 creation]."

Many former Israeli military and security officials also stepped forward to defend their country from any accusations of creating the malware. The former Israel Defense Forces (IDF) Brigade-General Pinchas Barel Buchris stated that no serious allegations could be made against Israel until more information would be released by Kaspersky Lab. The general wanted to know "what is in the footprint" of the Duqu 2.0 and "what it was compared to".

The former head of the Israel Security Agency Carmi Gillon took a different approach. He pointed out that there are more than enough other powerful nation-states, which could be interested in compromising the popular anti-virus software. He concluded that "it could be someone else", not necessary Israel.

The former IDF Brigade-General Yair Cohen, attending a cyber-conference at Tel-Aviv University, gave at least four more reasons for rejecting any possible connections between Israel and Duqu 2.0. The first argument was that "Kaspersky has its own interests" in that case. Unfortunately, there was no further explanation or clarification offered.

The second argument was that any similarities in the programming code of the newly discovered malware and Stuxnet could not be relevant after five years passed by. Anybody could use Stuxnet's code, that is why there was no reason to establish a possible connection of that spy-operation to Israel or the United States.

The third argument was related to costs associated with creating a sophisticated malware. The general pointed out that everybody knows that Stuxnet was created "with $100 million by a superpower". If that was the case, why did that "superpower" with unlimited resources not bother to change the programming code during those five years?

Saving the best for the last, Yair Cohen, the former head of the Unit 8200, could not resist to include in his last argument a sarcastic comment. More than likely, he was referring to the widely circulated statement of similarities between the digital signatures of Duqu 1.0 and

Duqu 2.0. Cohen started with a condition: "if in theory", it is possible that Duqu 2.0 was created by Israel or European Union or the United States, then one important piece of evidence is still missing. The former IDF Brigade-General said that there is "no sign of the name Shoshanna from Unit 8200 written" in the programming code of Duqu 2.0.

Shoshanna is a Hebrew feminine name. One of its American and European equivalents is the name of Susan. With that in mind, the statement about "Shoshanna from Unit 8200" had a special meaning. That was almost like saying: since nobody was able to show me a signature of Susan from NSA in the programming code of that nasty Stuxnet virus, it could not be the US, who had originated it.

Very soon other things than the cyber-spying began to dominate news cycles, and everything became quiet for the next two years. Until, in October 2017, the forgotten story about the Duqu 2.0 intrusion into Kaspersky Lab's very own computer network returned to the spotlight with a new twist.

It all started with an article published by The Wall Street Journal on October 5, 2017. In the article "Russian Hackers Stole NSA Data on U.S. Cyber Defense", Gordon Lubold and Shane Harris revealed that two years ago Russian hackers were able to access a National Security Agency (NSA) contractor's personal computer and steal some state-of-the-art hacking tools used by NSA. On his computer at home, the hacked NSA contractor used an anti-virus software made by Kaspersky Lab. According to the article, that software helped Russian hackers to identify and locate spying tools that belong to the NSA on that particular personal computer before extracting them. As the general public had learned two months later, the NSA contractor turned out to be an actual employee of the NSA.

That was just a beginning because the news kept coming. Just several days later, when the general public was still digesting the revelation that another NSA contractor, following examples of Edward Snowden and Harold Martin had been siphoning top government secrets into his personal computer, two more articles were published. On October 10, 2017, The New York Times came out with the article "How Israel caught Russian hackers scouring the world for U.S. secrets" by Nicole Perlroth and Scott Shane. In this article, the ultimate role of Israeli spies in cyber-attack on Kaspersky Lab in 2014 was revealed.

According to the mentioned above article in the New York Times, nobody else but Israel warned the U.S. about Russian spies, who were using Kaspersky Lab's software as some analog of Google search but strictly focused on looking for classified information. Unnamed multiple sources also told journalists that recently Russian hackers stole classified information from a home computer of an NSA employee, which used Kaspersky Lab's anti-virus.

It was a strange choice of computer security product because at work employees of the NSA are banned from using software made by Kaspersky Lab on their computers because of security concerns. It seems like those concerns had some grounds, which were proved by data collected during the Duqu 2.0 malware spy operation. Of course, the data itself remains classified but the reasoning of the NSA is understandable.

Since Kaspersky Lab anti-virus products have access to every single file on a computer, where they had been installed, four hundred million computers around the world had been sending information to servers in Moscow, where it was analyzed not only by information security analysts but possibly by Russian intelligence officers searching for foreign secrets.

One more day and more news became available. The Wall Street Journal published its very own revelation on the same subject – the article "Russia Has Turned Kaspersky Software into Tool for Spying" by Shane Harris and Gordon Lubold. The article confirmed that Israel gave a warning to its US counterparts. The warning was supported by screenshots and other information collected by the Duqu 2.0 malware inside the Kaspersky Lab corporate network.

Those revelations explained certain actions of the United States government taken against Kaspersky Lab, which took place earlier. Such as a decision by the General Services Administration (GSA) announced on July 17, 2017. On that day, GSA, the agency in charge of the centralized purchasing of goods and services for the government, removed Kaspersky Lab from the list of its authorized vendors. Or the announcement by the then-acting secretary of Homeland Security Elaine Duke. On September 13, 2017, she gave federal agencies ninety days to remove all Kaspersky Lab products from their computers.

Not all United States allies in Europe immediately fell in line with the US course of action against Kaspersky Lab. On October 11, 2017, the German federal cyber agency BSI was quoted by Reuters as saying, "There are no plans to warn against the use of Kaspersky products since the BSI has no evidence for misconduct by the company or weaknesses in its software."

As for the Israeli intelligence leadership, they were not thrilled to read about their classified cyber-operations in foreign publications by the New York Times and the Wall Street Journal. Their top-secret information, which had been routinely shared with the U.S. counterparts, was leaked to the press exposing Israeli capabilities in cyber operations. At the time of that exposure, several departments of Israeli governments continued using Kaspersky Lab anti-virus software.

Even more, the National Cyber Security Authority, an Israeli agency (during 2016-2018 in charge of protecting civilian networks from cyber-attacks), was closely cooperating with Kaspersky Lab. The National Cyber Security Authority provided the Russia-based company with information about malware, which Israelis had detected.

In the article "Israel's Kaspersky Hack Reveals That Russia Is Crossing All the Red Lines" by Anshel Pfeffer, the Haaretz Newspaper quoted Boaz Dolev, the CEO of ClearSky Cybersecurity, as saying, "there is an assumption in the cyber industry that many leading security companies in different countries cooperate with their governments. But what has been reported, that the internal search engine of the Kaspersky product was being used for targeted espionage in users' computers, is breaking all the red lines."

After all those revelations, now it was Kaspersky Lab's turn to deny at first and then explain everything.

Chapter 6.
Kaspersky Lab Explains Everything

The response to the Wall Street Journal article "Russian Hackers Stole NSA Data on U.S. Cyber Defense" from officials of Kaspersky Lab was immediate. Answering the requests for comments, they pointed out that the company had not received any proof of the accusations brought up in the article. Kaspersky Lab strongly denied any wrongdoing and having any inappropriate ties to the Russian government as well as any other governments. Also, it expressed its willingness to cooperate with the U.S. government during an investigation and asked for additional information from official sources. The responses by Kaspersky Lab were summarized in the press release dated October 5, 2017.

In his personal blog, Eugene Kaspersky, the CEO of Kaspersky Lab, expressed his own opinion. He stated that his company had been under continuous attacks by the U.S. media since the 1990s. That was the time when Kaspersky Lab first entered the North American market. Those attacks intensified at the beginning of the 2010s. The American media accused Kaspersky Lab of working for Russian intelligence without providing any proof or evidence because no such proof exists. Eugene Kaspersky concluded that besides he did not know the real reason behind the latest accusations, all that was paranoia, the bonfire of the Inquisition, and witch hunt.

During a phone interview with the Associated Press journalist Raphael Satter, on October 23, 2017, Eugene Kaspersky suddenly confirmed that at one point of time Kaspersky Lab had in its possession some files created by the Equation Group [the name given by Kaspersky Lab's analysts to the elite hacking team from the NSA]. That was in 2014 when a group of analysts entered his office. As Kaspersky recollected, "they told me that they have a problem." After learning about the nature of the problem, Kaspersky immediately said, "It must be deleted." When asked, if he informed the NSA about that accident, Eugene Kaspersky declined to answer.

Those official statements were shortly followed by the results of the internal investigation. On November 16, 2017, Kaspersky Lab released "Investigation Report for the September 2014 Equation malware detection incident in the US". At the very beginning of its report, Kaspersky Lab made a disclaimer that the investigation had been done "by using multiple analysts of non-Russian origin and working outside of Russia to avoid even potential accusations of influence". This was a somewhat strange statement because the investigation had been done by Kaspersky Lab's very own employees on the company payroll.

From the start of their investigation, Kaspersky Lab's analysts suspected that the incident must be somehow connected with the ongoing research of malware created by the Equation Group (which is widely believed to be a part of NSA's hacking operations). The digital signatures, produced by Kaspersky Lab to identify the Equation Group's computer viruses, contained names of hacking tools such as "Equestre", "Equation", "GrayFish", "Fanny", and "DoubleFantasy. Searching the company database for those digital signatures, the analysts found suspicious activity in September 2014, when the digital signature "HEUR:Trojan.Win32.Equestre.m" pulled out a large number of similarities on one computer.

The internal investigation concluded that Kaspersky antivirus software, installed by a user of the Version FIOS (Fiber Optic Service), in the Baltimore area, detected the digital signatures of malware originated by the Equation Group inside the user's computer. Between September 11, 2014, and November 17, 2014, the antivirus sent copies of detected files for future analysis to the Kaspersky Lab. The files ended up in the company's storage awaiting human attention.

Even after a brief look at those files, Kaspersky Lab's analyst realized that the files were found not on a victim's computer but the computer of the developer of the Equation Group hacking tools. Among the findings was a 7-Zip archive containing executable modules of some already known and previously unknown hacking tools from the arsenal of the Equation Group plus four documents with classification marks.

The investigation report made a special emphasis on the fact that the possibly classified documents were not intentionally detected during the search for malware, but they were picked up as the part of a compressed files archive, which was originally identified as infected by malware. It was unexpected that the infected archive turned out to be the malware itself, including previously non-discovered versions.

After that incident, Kaspersky Lab adopted a new policy "for all malware analysts which are required to delete any potential classified materials that have been accidentally collected during anti-malware research or received from a third party."

That 7-Zip archive, discovered by Kaspersky antivirus software, was later deleted at the direction of the company's CEO Eugene Kaspersky. The reason behind that decision was that for malware detection there is no need to keep all the files (especially ones that could contain possibly classified information). Only malware binaries were saved to create digital signatures for future detection.

After that step-by-step denial of its possible involvement in the story about stealing NSA data, Kaspersky Lab explained who and how could steal NSA's hacking tools. According to the same report, the user with NSA's programming codes in his personal procession decided to install a pirated version of Microsoft Office 2013. At 11:38 pm on October 4, 2014, he double-clicked the "setup.exe" file to activate the software. Inside that pirated file, there was hidden malware. The fact that Kaspersky Lab's software detected that malware as already running pointed out that the anti-virus was disabled during the download of the infected copy of Microsoft Office. Otherwise, Kaspersky antivirus would prevent the execution of malware.

The malware was identified by Kaspersky Lab as "Smoke Bot" (a.k.a. "Smoke Loader"). It was first created by a Russian hacker in 2011 and had been offered for sale to other hackers. At the time of the incident, that particular malware was trying to connect to its command and control server registered to a Chinese person by name Zhou Lou, based in Hunan province.

Overall, during the period between September 11, 2014, and November 17, 2014, the Kaspersky Lab software was able to detect

121 more computer viruses other than Equation Group malware inside the same computer. The report concluded that "the user's apparent need for cracked versions of Windows and Office, poor security practices, and improper handling of what appeared to be classified materials, it is possible that the user could have leaked information to many hands."

In its report, Kaspersky Lab reinforced its earlier statement that it never created digital signatures for its computer security products with names of classification markings [such as "secret", "top secret", "classified", etc.]. According to the company policy, after an analyst creates a digital signature, it goes for review by another group of analysts to ensure proper checks and balances. In addition to the internal procedures, there are external scrutinizing of digital signatures released by Kaspersky Lab.

The authors of the "Investigation Report for the September 2014 Equation malware detection incident in the US", stated,

"Considering that our signatures are regularly reversed by other researchers, competitors, and offensive research companies, if any morally questionable signatures ever existed it would have already been discovered. Our internal analysis and searching revealed no such signatures as well."

Kaspersky Lab's report, which was compiled in questions-and-answers format, also came up with an explanation of the possible origin of reported screenshots made by the Duqu 2.0 malware mentioned in the media. Traditionally the best was saved for the last. Answering the very last question "Assuming cyberspies were able to see screens of our [Kaspersky Lab's] analysts, what could they find on it and how could that be interpreted?" The answer started with the sound statement that, "We [Kaspersky Lab] have done a thorough search for keywords and classification markings in our signature databases. The result was negative: we never created any signatures on known classification markings". Then the authors of the report mentioned one possible incident, which could be wrongly interpreted as a creation of the digital signature searching for classification marks. The incident was related to the TeamSpy malware.

Kaspersky Lab announced the discovery of the TeamSpy malware on March 20, 2013. That malware allowed its originators to exploit the popular remote access software TeamViewer, so they could remotely control computers of their targets (government organizations, political and human rights activists predominantly located in former Soviet Union states and Eastern Europe). The malware would search for files with the extensions specific for Microsoft Office Word (*.doc), Excel (*.xls), Access (*mdb), Outlook (*pst), and PDF files (*.pdf). At the same time, the malware was looking for file names that contain particular keywords such as "pass", "secret", "секрет" [written in Cyrillic alphabet this word translates from Russian as "secret"], "saidumlo" [which means "secret" in Georgian, when written in Latin alphabet] and few others.

Since those keywords could be used to detect the malware, back in 2015, a Kaspersky Lab's analyst compiled a signature containing keywords "*saidumlo *, *secret*.*, *xls, *.pdf, *.pgp,*pass*.*". To avoid false positives, the analyst added a path to the folder ProgramData\Adobe\AdobeARM, used by the malware. If it would be safe to assume that the Duqu 2.0 malware was able to record the creation of this digital signature, it could produce "exposing" screenshots. As the authors of the report wrote, "Despite the intentions of the malware analyst, they could have been interpreted wrongly and used to create false allegations against us, supported by screenshots displaying these or similar strings."

Even though the report by Kaspersky Lab gives the plausible explanation of classified screenshots collected by Israeli intelligence and shared with its US counterparts, the timing of events is questionable. Kaspersky Lab published its report of the TeamSpy malware on March 20, 2013. It means that at that time digital signatures of the malware had been already written and Kaspersky antivirus software was updated to protect its customers. In 2015, for an unknown reason, the Kaspersky analyst decided to create a new digital signature for the old malware. Unfortunately, the report did not clarify a reasoning or urgent need behind rewriting the signature two years later.

Chapter 7.
New Details Related to
the Incident of Stolen NSA's
Hacking Tools Had Emerged

Two weeks after the release of Kaspersky
Lab's "Investigation Report for the September
2014 Equation malware detection incident in the
US", new details had emerged. As it was
originally reported by The Wall Street Journal,
back in 2015, Russian hackers stole National
Security Agency (NSA) programming codes,
which were stored in a personal computer of
NSA's contractor. Later Kaspersky Lab corrected

the date of the incident by giving an approximate date as October 2014.

It turns out that not only the date of the incident was wrong in the original reporting but also the occupation of the victim. The hacked NSA contractor was an actual employee of the NSA by the name of Nghia Hoang Pho. He was not an ordinary employee but a high-skilled programmer, who worked in the Tailored Access Operations (TAO) unit. Since then, TAO changed its name to CNO (Computer Network Operations) but not the nature of its activities that was first exposed by Edward Snowden. CNO continues to develop and use hacking tools to penetrate foreign computer networks.

Nghia Hoang Pho, a Vietnamese-American, was hired by the NSA in 2006. During five years, until his arrest in March 2015, he was bringing home from his work at the NSA copies of classified materials in hardcopy and digital form. Some of the materials found during the search of his residence had a "Top Secret" classification. The security clearance originally possessed by

Nghia Hoang Pho did not authorize him to access classified information outside of work.

Despite the internal policy of not using foreign-made anti-virus software on the computers inside the NSA's networks, Nghia Hoang Pho decided to install Russian-made Kaspersky Lab software on his personal computer at home. Even worse, on October 4, 2014, he disabled it to install a pirated copy of Microsoft Office software. That software was infected with malware, which created a "back-door" for hackers.

After that, the story unfolded as it was described in The New York Times article "How Israel caught Russian hackers scouring the world for U.S. secrets" from October 10, 2017. As Nicole Perlroth and Scott Shane wrote, "Israeli intelligence officers looked on in real time as Russian government hackers searched computers around the world for the code names of American intelligence programs."

Since all the information gathered during the infamous Israeli operation against one of the biggest names in the information security industry remains classified, the quotation above could be questionable in part of scope and time. What is known, is that when in 2015, the Duqu 2.0 malware infiltrated the Kaspersky Lab network, Israeli intelligence operatives were able to find certain signs of NSA's data. They shared their discovery with the NSA, which in turn started its very own investigation under the code-name Red Magic. The investigation ended up with an arrest of NSA's programming code developer Nghia Hoang Pho.

The former NSA employee used to live in Ellicott City, Maryland, twelve miles west of Baltimore. The location of his home matched the location provided by Kaspersky Lab in its report, which described then-unknown Pho as a user of the Verizon FIOS (Fiber Optic Service), in the Baltimore area. Nghia Hoang Pho used to work at NSA's headquarters in Fort Meade, Maryland, approximately 15 miles south from his home.

On September 25, 2018, Nghia Hoang Pho was sentenced to five and a half years in prison for unauthorized and willful retention of classified materials by the federal court in Baltimore, Maryland. The US government attorney Robert Hur gave the following statement, "As a result of his actions, Pho compromised some of our country's most closely held types of intelligence, and forced NSA to abandon important initiatives to protect itself and its operational capabilities, at great economic and operational cost."

The Revenge of Malware

Chapter 8.

The Brief Biography
of Eugene Kaspersky,
According to Bloomberg News
and Others

No matter who is writing about Kaspersky Lab, they would always mention a few selected facts from the biography of the company founder and CEO – Eugene Kaspersky. Those facts related to Kaspersky's education, his military service during the times of the Soviet Union, and connections to Russian security services. Take for example the article "Russia's Top Cyber

Sleuth Foils US Spies, Helps Kremlin Pals" at Wired magazine published in 2012.

In this article, Noah Shachtman described the biography of Kaspersky with its major milestones. At the age of 16, Eugene began his studies at the Institute of Cryptography, Telecommunications, and Computer Science (IKSI), which was an educational wing of KGB [Soviet security service – the equivalent of FBI]. After his graduation in 1987, Kaspersky served as "an intelligence officer in the Soviet Army". In 1991, his former instructor from the institute helped Eugene Kaspersky with his early discharge from the army and hired him. In 1997, Kaspersky together with his wife Natalya and his friend Alexey De Mont De Rique launched their own computer security company. In 2011, after FSB [Russian Federal Security Service] helped to release the kidnapped twenty-year-old son of Eugene Kaspersky, that event cemented even stronger the close cooperation between Kaspersky Lab's founder and Russian security services.

In 2013, a journalist Lorenzo Franceschi-Bicchierai asked Eugene Kaspersky about the article in Wired magazine, which Franceschi-Bicchierai helped to prepare as an intern. Kaspersky replied that, in his opinion, his competitor Symantec paid for the article "Russia's Top Cyber Sleuth Foils US Spies, Helps Kremlin Pals". In his article "Who's Afraid of Kaspersky?", Franceschi-Bicchierai wrote how upset about the Wired magazine article Kaspersky was by pointing out to the fate of Paul Roberts. Roberts, who was at that time an editor of ThreatPost (the blog about cyber-security issues), was fired after he retweeted the link for that article. As it turned out, ThreatPost had been unofficially fully funded and controlled by Kaspersky Lab, which ordered to ignore the article. Paul Roberts did not follow the directive and paid the consequences.

It is easy to understand why Eugene Kaspersky was so upset. From reading the article, readers would have the impression that Kaspersky spent plenty of hours with the journalist, who was following Eugene around everywhere, from Kaspersky Lab's headquarters to his private quarters inside a gated community in Moscow, drinking expensive whiskey and eating homemade blintzes with caviar. Kaspersky even let the journalist look at his top-secret project room, where even Kaspersky's personal assistant was not allowed to enter. The journalist took a glimpse in Kaspersky's closet, where he spotted a carefully preserved military uniform from the owner's times in the Soviet Army. Perhaps after such fine wining-and dining, Kaspersky expected a more friendly written article then he received.

Bloomberg News wrote about Eugene Kaspersky in a very similar way as the Wired magazine. On March 19, 2015, Bloomberg News came up with the article "The Company Securing Your Internet Has Close Ties to Russian Spies" by Carol Matlack, Michael Riley, and Jordan Robertson. The article stated that Kaspersky Lab has been focused on exposing cyber-spying activities by the U.S.A., United Kingdom, and

Israel while paying little attention to the same activities by the Russian Federation. The explanation, offered by the journalists, was that the CEO of Kaspersky Lab - Eugene Kaspersky had developed long term ties with Russian intelligence agencies. Those ties began during the times of Soviet Union with his work for KGB [the Soviet security service – the equivalent of FBI] and had continued with his weekly sauna visits with friends, many of whom are Russian intelligence officials.

The special emphasis in the article was made on the fact that since 2012 Kaspersky Lab went through some drastic changes. The company suddenly stopped its business relationship with the US company - General Atlantic, based in Greenwich, Connecticut. The American partner was involved in preparations for filing an IPO (Initial Public Offer) for Kaspersky Lab. The change of course from becoming a publicly traded company to being a closely controlled private company was never properly explained. The other major change happened during the next two years, when most of the top managers, including all foreigners, had been replaced with Russians well-connected with FSB.

Eugene Kaspersky swiftly reacted to the Bloomberg article by posting "A Practical Guide to Making Up a Sensation" on his personal blog on March 20, 2015. In his post, Kaspersky approached the article the same way as he would analyses malware. Line by line, he quoted the statements from the article and provided his counterarguments.

Answering to the accusations of not investigating Russian cyber-spies, Kaspersky presented a list of ten reports related to malware (including Red October, Black Energy, and TeamSpy) attributed by some analysts to Russian intelligence. As for his work for KGB, Eugene Kaspersky wrote, "I've NEVER worked for the KGB... I studied mathematics at a school sponsored by the Ministry of Atomic Energy, the Ministry of Defense, the Soviet Space Agency and the KGB. After graduating, I worked for the

Ministry of Defense as a software engineer for several years". Kaspersky pointed out that the journalists were referring mainly to former disgruntled employees, looking for revenge.

Of course, Kaspersky gave his very own interpretation of his sauna visits. He wrote that "sometimes I do go to the banya (sauna) with my colleagues. It is not impossible that there might be Russian intelligence officials visiting the same building simultaneously with me but I don't know them." At the end of his post, Eugene Kaspersky thanked Bloomberg and its journalists who, in his own words, "performed a full system scan – and found nothing."

Nevertheless, about three years later all the statements made by Bloomberg in March 2015, had surfaced again. This time it wasn't Bloomberg. On January 22, 2018, the US-based BuzzFeedNews and the Russian-language newspaper Medusa in Latvia simultaneously published the article "Inside the Fight for The Soul of Kaspersky Lab" by Ilya Zhegulev. The article added more details to the story about

Kaspersky son's kidnapping and the power struggle between Kaspersky Lab's tech-managers united with a group of Western top-managers and newcomers, connected to FSB. Especially for those newly hired managers, a brand-new department was created. The main focus of the department was on providing technical assistance to FSB. The funny thing was that the name of the department "The Computer Incident Investigation Department" had some reference to J. R. R. Tolkien's "The Lord of the Rings". If abbreviated, the department's name in Russian spells as ORKI, which is the Russian transliteration of "orc." According to the article, people from the ORKI lived up to their nickname "orc" and soon took over the control of Kaspersky Lab, which resulted in the refusal of becoming a publicly traded company and followed by drastic changes in top management.

Back in 2015, the exchange of arguments between Bloomberg and Kaspersky attracted the attention of other media outlets. Among them was National Public Radio (NPR). On August 10, 2015, NPR aired its very own piece "Kaspersky Lab: Based in Russia, Doing Cybersecurity in the West" by Corey Flintoff. At first, listeners heard familiar talking points about Kaspersky's education at KGB sponsored school and his work at a military institute. Then NPR reiterated statements made by Kaspersky in his response to the Bloomberg article "The Company Securing Your Internet Has Close Ties to Russian Spies". After that came a surprising twist. [Keep in mind that Kaspersky Lab is one of NPR's sponsors].Suddenly, NPR quoted Russian investigative journalist Andrey Soldatov, an editor of Agentura.ru website, as saying that Kaspersky Lab "became more and more associated with Russian security services, helping them to catch some cyber criminals, helping to provide security for some very important projects", including cyber-security of Olympic Games in Sochi in 2014.

By itself, the fact of cooperation between governments and domestic cyber-security companies is nothing new. Even in the Bloomberg article, the journalists gave an example of American cyber-security company FireEye, which is financed by the CIA. FireEye published their findings related to Chinese and Russian hackers but nothing yet about American cyber-spies.

As if often happens in the world of news media, some special topics never fade. They may cool down for a while before becoming hot again. The topic of close relationships between Kaspersky Lab and Russian security services was one of those special topics. The new round of confrontation between Kaspersky Lab and Bloomberg News resumed in the summer of 2017.

Chapter 9.
Controversies Surrounding
Kaspersky Lab

In the United States, the end of the spring
and the summer of 2017 did not go well for
Kaspersky Lab. From a bystander's point of
view, events unfolded too fast and without
enough logical explanation. Rough rhetoric in
the U.S. Senate was followed by proposing some
drastic legislation directly affecting Kaspersky
Lab's business interests in the USA. Media
outlets were trying to explain everything but
every time it was too little and too late.

On May 11, 2017, the U.S. Senate Select

Committee on Intelligence had its annual Worldwide Threats hearing. During that open hearing, Kaspersky Lab suddenly became a subject for discussion. It all started with the United States Senator Marco Rubio from Florida. He addressed his question to all six directors of U.S. intelligence organizations, "Would any of you be comfortable with Kaspersky Lab's software on your computers?" All of them answered, "No."

The United States Senator Joe Manchin from West Virginia went further with his question, "Has it come with your IT people coming to you or have you gone directly to them making sure that you have no interaction with KL [Kaspersky Lab] or any of the contractors you do business with?" After receiving a negative answer from everybody from the heads of the intelligence community, Senator Manchin requested from them a verification that no contractor uses Kaspersky Lab software or any type of hardware with Kaspersky Lab software installed. Still, no details supporting such a great deal of concern from the U.S. senators were offered during that open hearing.

Eugene Kaspersky responded immediately. He used the "Ask Me Anything" forum on Reddit to reject all allegations. Also, he did not miss a chance to brag about his company continuing success by saying, "I respectfully disagree with their opinion, and I'm very sorry these gentlemen can't use the best software on the market because of political reasons." After that heated exchange, things get quiet but not for long.

On June 27, 2017, FBI agents showed up at the homes of Kaspersky Lab's employees in the U.S. No searches had been conducted. No arrests had been made. Just brief interviews. Nevertheless, the very next day the Senate Armed Services Committee included in its defense spending policy bill an unusual amendment. The amendment explicitly prohibited the Department of Defense from using Kaspersky Lab's software. The official title of the amendment was "Prohibition on Use of

Products and Services Developed or Provided by Kaspersky Lab". The author of the amendment, the United States Senator from New Hampshire Jeanne Shaheen, was quoted by Reuters as saying, "...ties between Kaspersky Lab and the Kremlin are very alarming." As always, no additional details or proof of Kaspersky Lab's involvement with the Russian government were released.

Again, it was Kaspersky Lab's turn to reassure everybody: no illegal connections between Kaspersky Lab and Russian intelligence ever existed. On July 1, 2017, during his interview with the Associated Press, Eugene Kaspersky did his best to try to address concerns of US lawmakers and prove his company's willingness to cooperate. He said, "Anything I can do to prove that we don't behave maliciously I will do it." To prove that he was serious, the CEO of Kaspersky Lab made the very unusual step of offering to give away his company's most important proprietary information, "If the United States needs, we can disclose the source code". He also offered his testimony in front of the U.S. Congress.

Not all of the information security experts were impressed with Kaspersky's bold offer to open the programming code of Kaspersky Lab anti-virus for audit by the U.S. government. In the blog post "Why a Kaspersky code audit doesn't really ensure security", Rendition Infosec's blogger pointed out that with any software using remote update feature, including Kaspersky Lab's anti-virus, code is constantly changed. What was audited today would not be the same tomorrow. In Rendition Infosec's opinion, Kaspersky's statement was no more than a publicity stunt.

At the same time, Rendition Infosec made it clear that it would like to see proof of Kaspersky Lab's wrongdoing. "Rendition encourages a thorough discussion on the topic with appropriate levels of disclosure to back claims that Kaspersky software poses a bona fide [real] threat to DoD [Department of Defense] networks."

Robert M. Lee, CEO of Dragos, expressed a similar opinion on Twitter, "The folks there I've worked with have tirelessly run down adversary campaigns, provided some of the best analysis on malware... If it's so important then tell the American people and allies. Or realize it's a paranoid and awful precedent to make."

Russian Federation weighed in to protect Kaspersky Lab's interests in the U.S. As Bloomberg reported on June 30, 2017, Russia's Communications Minister Nikolay Nikiforov warned about possible retaliation in case of any unilateral sanctions against the Russian company. The minister reminded that the Russian government uses a big quantity of American software and hardware, but he did not provide any details about which American software products could be banned.

Since the U.S. government had not released any classified information related to the connections between Kaspersky Lab and Russian intelligence, journalists could only count on leaks from unnamed sources speaking under the condition of anonymity and rely on their very own investigations. That was exactly what journalists from Bloomberg did.

The Revenge of Malware

Chapter 10.

Kaspersky Lab Returns to Spotlight

On July 11, 2017, Bloomberg News came out not with one but with two articles exploring an already familiar topic – close ties between Kaspersky Lab and Russian intelligence and security services. One article, written by Jordan Robertson and Aki Ito, had the title "Why U.S. Officials Are Worried About This Russian Firm". The title of the other article, written by Jordan Robertson and Michael Riley, sounded like an answer to the question asked in the first article - "Kaspersky Lab Has Been Working with Russian Intelligence".

Traditionally articles about Kaspersky Lab have been based on unnamed sources sharing classified information. This time, the Bloomberg journalists had been able to show some proof. Previously unreported emails between Eugene Kaspersky and the senior staff of Kaspersky Lab, dated back to 2009, somehow ended up in the hands of the journalists. The emails were a part of the discussion related to the new at the time Kaspersky Lab's project – protection from distributed denial-of-service (DDoS) attacks.

The goal of any DDoS attack is to force a website (or to be more precise a server that supports the website) out of service by overloading it with fake service requests. Since the server can simultaneously process only a limited amount of service requests (based on its bandwidth limitations), attackers prepare their attack by gaining control over a significant number of computers or Internet-connected devices such as refrigerators, washers or other representatives of the Internet-of-things. Those hacked devices form a botnet (the combination of words roBOT + NETwork).

After the army of botnets receives a command to attack, it starts continuously sending requests to just one website creating a DDoS attack. The server is overwhelmed by the huge number of requests, and as the result crashes or begins to ignore legitimate requests. This is how the situation of denial of service is created for regular users or customers, who now cannot access the website.

In the above-mentioned internal email correspondence between Kaspersky Lab's officials, the main subject was a secret project by Kaspersky Lab to develop software, which would protect its customers from DDoS attacks. According to the journalist's interpretation of the emails, Kaspersky Lab designed the software per request of FSB (the Russian analog of FBI). The scope of the finished project went far beyond just a technological side (software by itself). It expanded to include Kaspersky Lab cooperation with Internet providers to locate originators of cyber-attacks, and as Eugene Kaspersky phrased it - "...active countermeasures (about which, we keep quiet) and so on".

No surprise that the phrase "active countermeasures" has caught the journalists' attention. As they explained in the article, this is "...a term of art among security professionals, often referring to hacking the hackers, or shutting down their computers with malware or other tricks." As an unnamed source told the journalists, Kaspersky Lab does much more than that for Russian security services. It provides them with real-time information about hacker's whereabouts and even goes with them to the hacker's location. More than likely it was a reference to "The Computer Incident Investigation Department" described later by BuzzFeedNews and the Russian-language newspaper Medusa.

As always, Kaspersky Lab immediately reacted with its very own interpretation of leaked emails. It posted the statement "Kaspersky Lab response clarifying the inaccurate statements published in a Bloomberg Businessweek article on July 11, 2017". In its statement, Kaspersky Lab insisted that the Bloomberg article misinterpreted the content of emails attributed to Kaspersky Lab and manipulated publicly known facts.

As for the software that protects customers from DDoS attacks, Kaspersky Lab stated that it was developed to fulfill existing market demand but not at the request from any of Russian security services, including FSB, which also never had been a user of the DDoS protection software.

When it comes to cooperation with Internet providers to locate originators of cyber-attacks, Kaspersky Lab explained that it "does not cooperate with hosting companies to locate bad actors". At the same time, if Kaspersky Lab's experts could recognize an origination of DDoS attack from a specific data center, they would pass this information to hosting companies. Then hosting companies would have a chance to stop a DDoS attack.

In Kaspersky Lab's view, the term "active countermeasures" was misinterpreted. It meant "the DDoS intelligence system, which alerts that there is an emerging DDoS-attack against a customer through monitoring the activity of DDoS botnets." Kaspersky Lab rejected as false Bloomberg's interpretation of "active countermeasures" as "hacking the hackers" because this type of action is illegal. Instead, Kaspersky Lab insisted that it had been cooperating with law enforcement organizations, including Interpol, in shutting-down botnets by providing technical support and expertise.

Kaspersky Lab was not alone in disagreeing with the use of the term "active countermeasures" in the context of Bloomberg's article. In the blog post "Honestly Evaluating the Kaspersky Debate", Rendition Infosec made it very clear that there is no such standard term in the information security industry. It gave the following interpretation, "We know of no such standard definition for "active countermeasures." Even if Bloomberg got this definition from an infosec expert, any expert worth quoting would have told Bloomberg that their definition was one of many and not "generally accepted" by the community. That this wasn't reported makes the whole article reek of bias – where there's smoke, there's usually fire."

Arguably, the most intriguing part of Kaspersky Lab's statement was its response to the accusation that it supplies Russian security services with the "real-time information about hacker's whereabouts". Kaspersky Lab stated that it "doesn't provide any government agencies, nor other parties, with information on location of people and doesn't gather "identifying data from customers' computers" because it is technically impossible."

That "technically impossible" phrase, used by Kaspersky Lab, did not sit well with the information security community. Rendition Infosec reacted to that statement in the blog post "Honestly Evaluating the Kaspersky Debate". In its professional opinion, for any anti-virus software maker to say that it is "technically impossible" to collect information, which could be used to identify end-user would be knowingly making a false statement.

As Rendition Infosec explained, anti-virus software uses telemetry to gather and submit various data from a user's computer to its headquarters. That data is used to identify malicious activities of computer viruses. An important part of routinely collected data is information stored in registry keys.

Registry keys are folders inside Microsoft Windows operating system's database called registry. When a new software has been installed on the computer, this new program stores its start-up files in a registry key. After computer restarts, the software needs to access those files to reactivate itself. Malware uses registry keys the same way as any legitimate software to survive the computer's reboot. Since malware could store its configuration and exfiltration files in registry keys, anti-virus software must have access to all registry keys.

Among other data, which could be found in registry keys, there are computer's name, usernames, email address associated with Microsoft account, name of Wi-Fi network, URLs of recently visited websites, computer's processor serial number, Windows software unique ID, Kaspersky anti-virus unique ID, and much more.

Based on everything mentioned above, Rendition Infosec concluded, "Any centrally managed antivirus that can't collect this sort of telemetry isn't doing the best possible job for its customers. Whether anyone at Kaspersky has used the capability to assist intelligence or law enforcement, the capabilities almost certainly exist – and not just in Kaspersky software."

Overall, Rendition Infosec reflected the prevailing opinion in the industry. In the case of continuous allegations of Kaspersky Lab having alarming connections with the Russian government and intelligence organizations, both the U.S. government and media still did not present any meaningful facts supporting their claims.

Chapter 11.
Smoke and Mirrors

On October 10, 2017, the public finally had a rare chance to glimpse behind "smoke and mirrors" surrounding the U.S. government hostility against Kaspersky Lab. The article "The confrontation that fueled the fallout between Kaspersky and the U.S. government" by Patrick Howell O'Neill was published by CyberScoop. According to the article, things between Kaspersky Lab and the U.S. government went wrong even before the Duqu 2.0 malware was discovered inside Kaspersky Lab's corporate network, which was made public on June 10, 2015.

During the first half of 2015, sales representatives of Kaspersky Lab were very aggressive in their efforts to sell their company software to U.S. intelligence and law enforcement agencies. One part of their sales pitch attracted the attention of FBI experts, who worked for the Counterterrorism Division. They were intrigued by the statement of Kaspersky Lab's representatives that their anti-virus software could be leveraged in a way, which would assist with the capture of terrorist-related targets in the Middle East. For FBI experts the true meaning of that statement was that the Kaspersky Lab's product could be used for spying. Spying not only in the Middle East but in other regions as well, including North America.

Still, other divisions of the FBI, at least for the time being, chose to ignore "red flags" and became serious about cooperation with Kaspersky Lab. As a part of the "due diligence" process, FBI agents paid home-visits to employees of the company on June 27, 2015. During the interviews, they asked questions about the relationship between the U.S. offices of Kaspersky Lab and Moscow headquarters.

The growing interest from the FBI toward Kaspersky Lab had been spotted not only by media, who reported that FBI agents questioned U.S. employees of Kaspersky Lab but also by Russian security services. In July 2015, during a meeting in Moscow between FSB and CIA officials, Russians presented a diplomatic demarche to their U.S. counterparts, urging them to leave Kaspersky Lab alone. Of course, the original document, which has not been made public, used more formal language such as "malicious interference" against Russia-based company.

As an unnamed U.S. official explained CyberScoop, "This was a clear signal from the FSB to the U.S. to get off their intelligence asset. If this was from the foreign ministry, that would have been different. It is extremely rare and a different message when an intelligence agency démarches you." Alarmed by that unusual and very strong reaction from Russia, the FBI started its very own counterintelligence operation focused on Kaspersky Lab's ties with the Russian government and intelligence services.

Even though anonymous sources told CyberScoop that turning point in the relationship between the U.S. government and Russian cyber-security company was the diplomatic demarche, don't forget that the summer of 2015 had one more relevant event. The demarche took place in July but one month earlier, on June 10, Kaspersky Lab publicly admitted that it was targeted by state-of-the-art malware from the Stuxnet family.

The Duqu 2.0 malware was exposed, added to the virus database, and attributed to Israel. Since the secret operation by Israeli intelligence against Kaspersky Lab was terminated, it is safe to assume that around the same time the exchange of collected intelligence took place. Articles in the New York Times and the Wall Street Journal, published in 2017, gave readers some general idea about the data shared by Israelis with the NSA. Kaspersky Lab's anti-virus software had been presented as a "Google search" for Russian spies and a tool for extracting top secrets from computers around the world.

More than likely, the NSA briefed on that subject the U.S. government and other members of the intelligence community. That is how the FBI received some missing pieces for its investigation. That would explain the next fact described by CyberScoop. At the end of 2015, the FBI was actively advising the major U.S. companies to stop using Kaspersky Lab's software. At the same time, the FBI presented a classified intelligence report to Congress As for the public, as always, it got nothing, except leaks to the media.

The Revenge of Malware

Chapter 12.
The NSA and Kaspersky Lab:
Rocky Relationships

There should be no surprise that relationships between the NSA and Kaspersky Lab could never be good. They are in different businesses. The NSA is in the business of spying, including signals intelligence operations in cyber-space, and Kaspersky Lab is in the business of defending its customers from cyber-intruders, including nation-state hackers. Even more, according to Edward Snowden's leaks, the anti-virus software produced by Kaspersky Lab became a big obstacle for intelligence operations.

For that reason, Kaspersky Lab had been one of the prime targets of the NSA, at least since 2008.

In its turn, Kaspersky Lab publicly announced the presence of the NSA in the cyber world as a powerful threat actor under the name - Equation Group. On February 16, 2015, Kaspersky Lab issued a press release titled "Equation Group: The Crown Creator of Cyber-Espionage". The press release named the Equation Group as the most powerful threat actor among more than sixty threat actors, which Kaspersky Lab has been monitoring. The Equation Group used the most sophisticated and powerful cyber-tools, and it was operating them through more than three hundred command and control servers around the world.

One of the most intriguing announcements from Kaspersky Lab's press release was a statement establishing a connection between the Equation Group malware – the computer worm Fanny and previously discovered Stuxnet and Flame. Turns

out that two zero-day vulnerabilities, used by
Stuxnet in 2009 and 2010 (one of them was
copied directly from Flame's programming code
to Stuxnet's code) were not completely new at
the time. Those zero-day vulnerabilities were
originally part of Fanny's bag of tricks and were
first put in use as early as 2008.

Fanny and Stuxnet had many more things
in common than shared zero-day vulnerabilities.
Both were developed to penetrate air-gapped
computer networks, which for security reasons
are never connected to the Internet. Even the
way Stuxnet and Fanny spread around was
similar. Fanny infected computers through USB
memory sticks, granted itself the highest level of
privileges and collected information about
computers within the network. That is why
Fanny and Stuxnet needed in the same zero-day
vulnerabilities. Of course, they had different
objectives. Fanny was a spy used to create maps
of air-gapped computer networks, and Stuxnet
had a mission to find and destroy uranium
enrichment centrifuges.

Of course, Kaspersky Lab could not avoid some bragging. For obvious reasons, Kaspersky Lab could not claim that it was the first to discover Stuxnet, even though the company later hired a person, who was the first to report Stuxnet. That is why, in this case, Kaspersky lab settled for less. It pointed out that Kaspersky Lab's anti-virus software detected Fanny and labeled it as malware way back in December 2008, only six months after the Fanny computer worm was created.

In the press release, Kaspersky Lab made a conclusion about the existence of "solid links indicating that the Equation group has interacted with other powerful groups, such as the Stuxnet and Flame operators – generally from a position of superiority. The Equation group had access to zero-days before they were used by Stuxnet and Flame, and, at some point, they shared exploits with others."

In another document ("Equation Group: Questions and Answers"), Kaspersky Lab went even further by stating that " the similar type of usage of both exploits together in different computer worms, at around the same time, indicates that the *Equation* group and the Stuxnet developers are either the same or working closely together." By 2015, there were little doubts about the origin of the Stuxnet computer worm. For this very reason, even though Kaspersky Lab did not say openly that the Equation Group is, in fact, the National Security Agency of the United States of America (NSA), everybody could easily guess about that.

Understandably, the public revelation of the existence of the Equation Group and its activities could not help in the improvement of the relationships between the NSA and Kaspersky Lab. Still, one year later, researchers from Kaspersky Lab did something to the NSA, which may be considered as a big favor. In 2016, they helped to expose a former NSA's contractor Harold T. Martin III. For twenty years, Martin had been stealing classified information from the NSA. By the time of his arrest on August 27, 2016, he accumulated fifty terabytes of secret data. His collection included some state-of-the-art malware.

As Kim Zetter explained in her article "How a Russian firm helped catch an alleged NSA data thief" published by POLITICO, Harold T. Martin III sent five private messages via anonymous Twitter account to employees of Kaspersky Lab. Those cryptic messages could be interpreted as a limited time offer of something of value. The first message was, "So... Figure out how we talk. With Yevgeny present." It followed up by the second message, "Shelf life, three weeks." Obviously, "Yevgeny" as an alternative spelling for the Russian name Eugene was a reference to the CEO of Kaspersky Lab Eugene Kaspersky.

Even more interesting was the fact that the first messages were sent on August 13, 2016, just thirty minutes before a hacker group under the name Shadow Brokers announced via Twitter the "Equation Group Cyber Weapons Auction". The announcement later followed up with samples of NSA's hacking tools. That coincidence made Kaspersky Lab's researchers suspect that Martin was a source of information for Shadow Brokers or Shadow Brokers himself.

Employees of Kaspersky Lab performed some online searches and quickly found out that the same username was used in communicating with them through Twitter account and for registration on one of the dating sites. The dating site had a picture of Martin himself plus some additional information about him. That information helped to find a LinkedIn account of Hal Martin. On August 22, 2016, the collection of tweets together with the information related to the identity of their author was emailed by Kaspersky Lab's researcher to an NSA employee, whom he met at a conference.

Only five days later, on August 27, 2016, FBI agents together with the SWAT team surrounded the house of Harold Martin III. During the search, they discovered a huge amount of classified information. The FBI immediately arrested Martin. Since Shadow Brokers continued to release NSA's secrets after the arrest, it became clear that Martin was not hiding his identity under the name of Shadow Brokers. Edward Snowden was one of the first to link Shadow Brokers with Russian intelligence in a series of his tweets on August 16, 2016.

Events of the following year, 2017, demonstrated that after the arrest of Martin, relationships between Kaspersky Lab and the NSA did not improve even a little bit. This is how the former general counsel for the NSA, Stewart Baker, commented on that in his interview with POLITICO in 2019. He said that the actions of Kaspersky Lab were "...a goodwill gesture toward the U.S. government." At the same time, he said, "I'm sure the people at Kaspersky are feeling as though they did the right thing and it did them no good."

Three years after Kaspersky Lab announced its discovery of the Equation Group, another discovery had been made public in March 2018. This time Kaspersky Lab exposed Slingshot – the sophisticated cyber-spying malware, attributed by some outside information security experts to the NSA. Slingshot had been active in Africa and the Middle East from 2012 to February 2018. Kaspersky Lab's press release was named "Kaspersky Lab uncovers Slingshot, the spy that came in from the router." In this case, the title says it all. Slingshot attacked its targets through hacked public Wi-Fi routers commonly used at Internet-cafes, in particular, Mikrotik routers. The majority of victims were located in Kenya and Yemen, others were in Afghanistan, Libya, Congo, Jordan, Turkey, Iraq, Sudan, Somalia, and Tanzania.

When victims wanted to access the Internet, their computers were connecting to a router first. During that initial connection, computers downloaded a dynamic link library (DLL) from an infected router. Slingshot arrived inside the DLL and began to load its very own modules. Those modules were responsible for recording passwords, keyboard data, clipboard content, and making screenshots. To avoid detection, Slingshot worked in the most privileged and trusted mode of the Windows operating system – kernel mode. The Slingshot malware hid stolen data in an encrypted form on the computer's hard drive. Those files were invisible for the operating system, which considered them as free space. If the operating system wanted to use that "free" space during a defragmentation process, Slingshot prevented the relocation of other files into occupied space by disabling defragmentation process. These are only a few tricks from Slingshot's impressive arsenal of spying tools.

Kaspersky Lab has a tradition to make its discovery of malware public during its annual Security Analyst Summit (SAS). The discovery of the Equation Group was announced at SAS 2015. Slingshot was presented at SAS 2018. At first, both events looked similar but there was a big difference. According to Lorenzo Franceschi-Bicchierai, the NSA knew about the coming announcement of the existence of the Equation Group for almost twelve months in advance. For that reason alone, the NSA had plenty of time to do damage control. The Slingshot announcement caught the NSA by surprise. It was completely unexpected.

In his article "Who's Afraid of Kaspersky?", published by MOTHERBOARD on May 22, 2018, Franceschi-Bicchierai pointed out that Kaspersky Lab, by announcing the discovery of Slingshot, exposed an ongoing secret operation by the United States Joint Special Operations Command (JSOC) against terrorists associated with ISIS. As a result, the U.S. had no other choice than to scrap its compromised digital infrastructure and look for other ways to collect intelligence.

It is not unusual for cyber-security companies to share with the public their discoveries related to counterterrorism operations performed by nation-states. At the same time, they follow the unwritten policy of "responsible disclosure". It means that cyber-security companies would give suspected originators of counterterrorism malware advanced warning. A good example of "responsible disclosure" would be the discovery of the Regin malware in 2014. The malware, attributed to UK's Government Communications Headquarters (GCHQ), had been uncovered much earlier but both, Kaspersky Lab and Symantec, waited to expose it publicly in order to avoid interference with GCHQ's ongoing operations.

In the case of Slingshot, everything looked the opposite. As an information security expert at CrowdStrike Michael Rea noticed on Twitter, "... one can't possibly help but think this was either a calculated burning of CT [counter-terrorism] operations for PR [public relations] or retribution for the past year." Of course, Kaspersky Lab had insisted that there was no retaliation of any kind because the Russia-based company reports on all discovered malware regardless of suspected originators' national origin. The last part of the above statement is entirely true under one condition. Only the company's paid subscribers get it all. The rest of the public might never see all the information about the newest cyber-threats and hacker, which is published fully only in private reports.

It is hard to say, what was really behind the Slingshot discovery announcement: revenge or the long-standing public relation tradition, as it was the case with the Equation Group. Either or, it did not help to improve relationships between Kaspersky Lab and the NSA.

The Revenge of Malware

Chapter 13.
Ban and court fight

Differences between the U.S. government and Kaspersky Lab had been quickly moving to the legal field. As a quick reminder, on July 17, 2017, General Services Administration (GSA) removed Kaspersky Lab from the list of authorized vendors. On September 13, 2017, the Department of Homeland Security (DHS) issued a directive that gave federal agencies ninety days to remove all Kaspersky Lab software from their computers.

When direct attempts to improve relationships with the U.S. government had failed, Kaspersky Lab switched its focus on doing damage control. As a part of that process, on October 23, 2017, Kaspersky Lab announced the Global Transparency Initiative. Under the initiative, data collected from computers using Kaspersky Lab anti-virus would be transferred to servers in Switzerland; the compilation of anti-virus databases and software would also be done in Switzerland; several regional Transparency Centers would be open. Each of those Transparency Centers would become a place, where governments and the company's trusted partners can access information about Kaspersky Lab, including source code.

News about the Global Transparency Initiative did not change the mindset of U.S. politicians. On December 12, 2017, President Donald Trump signed into law the defense policy spending bill – the 2018 National Defense Authorization Act. Part of this bill was a ban on the use of Kaspersky Lab software within all branches of the U.S. government. One of the people celebrating the ban was its author, U.S. Senator Jeanne Shaheen. She said, "The case against Kaspersky is well-documented and deeply concerning. This law is long overdue." Again, the "well-documented" part of the statement raised eyebrows of the general public, who had not been given the slightest opportunity to read or see any of the actual documents. Kaspersky Lab found itself in a similar situation, and it went to court. In particular to the U.S. District Court for the District of Columbia, where Kaspersky Lab appealed the directive of DHS, which required removal of Kaspersky Lab's products from federal agencies computers.

On December 18, 2017, Kaspersky Lab issued an open letter. In the letter, Kaspersky Lab informed the public about its appeal and provided some explanation. From Kaspersky Lab's point of view, there was no adequate due process, because DHS did not provide any information related to unlawful activity by Kaspersky Lab. For that reason, the company could not defend itself by contesting that information. As a result, "DHS has harmed Kaspersky Lab's reputation and its commercial operations without any evidence of wrongdoing by the company." In the end, Kaspersky Lab offered its cooperation through the Global Transparency Initiative.

On February 12, 2018, Kaspersky Lab made the second attempt to challenge the U.S. government in court. The company filed a lawsuit claiming that the provision of the 2018 National Defense Authorization Act (NDAA) that banned Kaspersky Lab's software was unconstitutional. One of the main arguments had sounded very familiar: there was no evidence of wrongdoing, which was made available to the public.

On May 30, 2018, after reviewing both lawsuits, the United States District Judge Colleen Kollar-Kotelly dismissed Kaspersky Lab's lawsuits as being unsubstantial. From her point of view, the goal of the NDAA and the DHS ban was not to punish Kaspersky Lab but to remove a risk to national information security.

The existence of the risk was determined by the Department of Homeland Security (DHS) based on the following:

- Federal agencies have been using Kaspersky Lab software.

- Kaspersky Lab plans to expand its sales to the federal government.

- Kaspersky Lab anti-virus software, like any other similar product, has privileged access to computer files and operating system, which could be exploited by hackers.

- Kaspersky Lab software collects and transmits data directly to servers in Russia or accessible from Russia.

- Russia has been engaged in malicious cyber activities against U.S. government networks.

- Kaspersky Lab as an organization and its officials have ties to the Russian intelligence agencies.

- Kaspersky Lab is a Russian company, and, under Russian law, it is obligated to assist Russian intelligence agencies in many areas, including an interception of communications within Russian networks.

At the same time, DHS stated that it

issued its ban not because Kaspersky Lab was "disloyal or guilty of any wrongdoing" but in order to mitigate the risks. In the court's memorandum opinion, the judge also quoted then-acting secretary of Homeland Security Elaine Duke as saying that the ban was aimed to curb "the ability of the Russian government, whether acting on its own or through Kaspersky, to capitalize on access to federal information and information systems provided by Kaspersky-branded products."

Judge Colleen Kollar-Kotelly ruled that the ban of Kaspersky Lab products under the 2018 NDAA is constitutional, and it would go into effect on October 1, 2018.

Chapter 14.

Final thoughts and Things to Consider

The real reason behind the 2018 ban of Kaspersky Lab software is still unclear. There are too many possibilities: from being collateral damage from the geopolitical war between two global powers to the revenge of exposed threat actors to intrigues of competitors. Looking not for a reason but for an episode that triggered that unfortunate chain of events, which ultimately led to the ban, seems to be a little bit easier. There are two possible answers, both of them dated back to 2015.

The first possible answer has been described in Patrick Howell O'Neill's article "The confrontation that fueled the fallout between Kaspersky and the U.S. government" in CyberScoop. That was an unusual sales pitch from Kaspersky Lab's sales representatives when they made an emphasis that their anti-virus software could be used to track down terrorists in the Middle East. Sounds somewhat similar to the infamous Slingshot. Maybe not, since Slingshot was malware but not an anti-virus program. Only intended targets turned out to be surprisingly similar.

The second plausible answer comes from two articles published by The New York Times and The Wall Street Journal. Nicole Perlroth and Scott Shane wrote "How Israel caught Russian hackers scouring the world for U.S. secrets", and Shane Harris and Gordon Lubold titled their article "Russia Has Turned Kaspersky Software into Tool for Spying". Both articles were based on information (as always from unnamed sources) about data collected by the Duqu 2.0 malware inside Kaspersky Lab's corporate network.

More than likely the ban of Kaspersky Lab software was triggered not by a single event but by a series of events. It is possible, that the two previously mentioned answers should be combined. Any sales pitch is just words. Screenshots and stolen files could amount to evidence. Of course, if they are available to the public and not tampered with.

The ban itself did not hurt Kaspersky Lab too much, because the level of sales to the U.S. government before the ban (as of September 2017) was about fifty-four thousand dollars, which was less than one percent of Kaspersky Lab's annual sales in the U.S.A. The damage had been done to the reputation of Kaspersky Lab, especially in North America, which had affected the bottom line of the company finances. The ban became a culmination of undesirable reporting by Bloomberg News, The New York Times, The Wall Street Journal, and other media outlets. None of that could make Kaspersky Lab and its CEO happy.

According to the Kaspersky Lab's press release from February 19, 2019, the company revenue in 2018 reached $726 million, which is a 4% increase in comparison with the previous year. Sales in Russia and Europe went up by 6%. At the same time, sales in North America dropped by 25% due to, how the company phrased it, "the challenging geopolitical situation".

It seems like Eugene Kaspersky himself blamed his competitors for the negative coverage in the media. At least, according to a post on the personal blog on October 17, 2017, it was the first thought that crossed his mind. That thought popped up in Kaspersky's mind when Eugene was reading in the New York Times that Israelis, or to be more precise the Duqu 2.0 malware, caught Kaspersky's anti-virus software stealing classified data from the U.S.-based computer. Or look at another situation. As the journalist, Lorenzo Franceschi-Bicchierai, recalls his 2013 conversation with Eugene Kaspersky about the article in the Wired magazine. Then

Kaspersky was convinced that his U.S. competitor Symantec paid for the article "Russia's Top Cyber Sleuth Foils US Spies, Helps Kremlin Pals".

It appears as Eugene Kaspersky was not the only one, who was interested in finding out the role of Kaspersky Lab's competitors in bad publicity. As Raphael Satter from Associated Press reported, in 2018, someone under the name of Lucas Lambert arranged several meetings with experts in cyber-security, who had been critical of Kaspersky Lab. The formal agenda of the meetings was to invite the experts to a fictitious cyber-security conference with an initial offer of $10,000 for a speech at the conference. For unknown, at least for cyber-security experts, reason, the conversation kept returning from discussing details of the conference to one topic: was it one of Kaspersky Lab's competitors, who influenced, or even paid, the experts to give their negative comments to media.

The investigation by Associated Press

found out that a company, which Lucas Lambert claimed to represent, never existed in real life. It had only a virtual presence as a website. Lambert itself was identified as Aharon Almog-Assouline, a former officer of an Israeli intelligence organization. There were also speculations that Almog-Assouline was employed by an Israeli private intelligence company Black Cube.

Black Cube denied any connections with "Lambert"-Almog-Assouline and his operation aimed at finding out if competitors were behind the critics of Kaspersky Lab. Associated Press tried to reach Kaspersky Lab for comments, but the Russia-based cyber-security company refused to answer. Raphael Satter's articles "Undercover spy targeted Kaspersky Lab critics", "Mysterious operative haunted Kaspersky critics", and "Israeli firm Black Cube denies link to spy targeting Kaspersky Lab critics" were published on April 17, 2019.

It is very important to remember that Kaspersky Lab has been a target of the U.S. and

U.K. intelligence for a long time. Western spies named Kaspersky Lab's computer security software as one of the main obstacles for their efforts of collecting signals intelligence. On one hand, if the number of organizations and people using Kaspersky Lab's anti-virus would decline, the job of Western spies becomes easier. They would benefit from that. On the other hand, if Bloomberg, The New York Times, and The Wall Street Journal are correct, under the scenario, when more Internet users are switching to Kaspersky Lab then the job of Russian spies would become easier.

Rendition Infosec discussed potential problems with all, not just Kaspersky Lab anti-virus software in its blog a while ago, back in 2017. In the first post titled "Is your antivirus software part of your threat model? Maybe it should be...", it pointed out on threats posed by anti-virus software. To name just a few, they are the following: anti-virus program may allow certain malware sneak into a computer (if malware was originated by a friendly intelligence organization), steal usernames and passwords, copy, modify or even destroy files of particular interest, etc.

The second blog post titled "Should Antivirus software be part of your threat model?" took a more systematic approach. This time all threats from anti-virus software were divided into three categories: confidentiality, integrity, and availability. Any anti-virus software threatens the confidentiality of users because it accesses every single file on the computer. The threat to integrity is in anti-virus software ability to modify data inside the computer's files. The threat to availability lays in anti-virus software's unique ability to cease any ongoing process and block user's access to any other software inside the computer. The conclusion was simple: install anti-virus software only from vendors that you trust.

During the hearing before the U.S. House of Representatives Subcommittee on Oversight and Committee on Science, Space, and Technology on October 25, 2017, U.S.

representative Clay Higgins quoted a letter from Troy Newman, a managing partner at information security firm Cyber5. In his letter, Newman compared an installation of anti-virus products with vaccination. He explained that before installing "any effective security software, we must first expose the system, making all information vulnerable. The security software has full access to all input and output operations. Security software is fully imbedded in such a way that it has complete access to total—to the entire system."

At the same hearing, Sean Kanuck, Director of Future Conflict and Cyber Security International Institute for Strategic Studies, pointed out that the total access by anti-virus program to all files on the computer makes it a target for foreign intelligence and criminal hackers. They would try everything to exploit that privileged access. Kanuck concluded, "I do not believe there is any network or any product that is perfectly secure. It's all a risk management issue."

It would be difficult not to agree with the above statement. The infamous computer worm Stuxnet demonstrated an ability to inject itself into a scanning process of anti-virus software. That ability allowed Stuxnet to travel unnoticed in a search for a specific program used to operate uranium enrichment centrifuges.

After all, it is not uncommon for governments not to trust foreign companies, when it comes to the security of the government's computer networks. During the interview with the television network France 24, Admiral Dominique Riban, from the French national IT security agency ANSSI, explained that in order to protect classified information he would not trust foreign companies, including Russian and Americans. He added, "France doesn't have friends in the cyber world. We have enemies, and we have allies."

As for Kaspersky Lab, life continues after the ban. As a part of its continuous efforts of rebuilding trust, Kaspersky Lab opened its first Transparency Center in Zurich, Switzerland in

November 2018. The opening of the second Transparency Center followed shortly, on April 2, 2019, in Madrid, Spain. The third Transparency Center serving the Asia-Pacific region will be open in Malaysia in 2020.

On June 4, 2019, Kaspersky Lab announced its rebranding. The new name and the new logo had been revealed to the public. Today the global company has overgrown the old anti-virus lab, so the word "lab" disappeared from the name. Now it is simply Kaspersky, and it is not written using Greek alphabet letters anymore. The renewed company, under the name Kaspersky, promises its four hundred million users and two hundred seventy thousand corporate clients worldwide not simply "cybersecurity" but complete "cyber-immunity".

Unfortunately, the rebranding did not save the renewed Kaspersky company from another controversy. On August 15, 2019, German c't Magazine published an article by Ronald Eikenberg under the title "Kasper-Spy: Kaspersky Anti-Virus puts users at risk". When testing Kaspersky Lab anti-virus for the March edition of the magazine, the journalist noticed that Kaspersky Lab anti-virus was leaving its digital ID, so-called Universally Unique Identifier (UUID) on every website visited by a user without user's permission. It meant that every user of Kaspersky software could be tracked all the time, even in "private" or "incognito" mode, regardless of which browser was used.

As always, Kaspersky Lab downplayed the significance of the data leak, which affected all users of Kaspersky Lab anti-virus (approximately 400 hundred million people) starting with the 2016 edition. The security issue was patched on April 4, 2019, and described in the security advisory CVE-2019-8286 (Common Vulnerabilities and Exposures). Still, in Ronald Eikenberg's opinion, the company did not do enough to protect its customers. Kaspersky anti-virus stopped injecting the unique to each user ID, but it continues to leave the collective ID representing the version of Kaspersky anti-virus software installed on a user's computer. The journalist had insisted that knowing the version of anti-virus software would make it easier for hackers to attack a user's computer.

As for Duqu, it did not entirely disappear. It managed to get back to the spotlight in April 2019. This time its appearance went unnoticed by the general public but attracted the attention of the members of the information security industry. The relatively new cybersecurity company Chronicle (the subsidiary of Alphabet Inc., which is a parent company of Google) reported a discovery of a transitional version of the Duqu malware. Researchers named it – Duqu 1.5, because in their opinion that was an "overengineered" experimental version of what would later become the most sophisticated living-in-memory Duqu 2.0 malware.

As of today, Duqu continues to be the one and the only malware that had its revenge. Never before in a history of information security, the discovery of malicious software would lead to a government ban of anti-virus programs made by a cyber-security company, which exposed the malware.

The Revenge of Malware

Sources

Chapter 1.

1. Duqu is back: Kaspersky lab reveals cyberattack on its corporate network that also hit high-profile victims in - Western countries, the -Middle East and Asia. By Kaspersky Lab. Kaspersky Lab. June 10, 2015. https://www.kaspersky.com/about/press -releases/2015_duqu-is-back-kaspersky- lab-reveals-cyberattack-on-its-corporate- network-that-also-hit-high-profile-victims- in-western-countries-the-middle-east-and- asia. Accessed on January 28, 2019.
2. 70th anniversary of liberation of the Nazi German concentration and extermination camp Auschwitz-Birkenau. Auschwitz- Birkenau State Museum. January 27, 2015. http://70.auschwitz.org/index.php?option =com_content&view=article&id=18&Itemid =134&lang=en Accessed on January 28,

2019.

3. Stepson of Stuxnet stalked Kaspersky for months, tapped Iran nuke talks. By Dan Goodin. Ars Technica. June 10, 2015. https://arstechnica.com/information-technology/2015/06/stepson-of-stuxnet-stalked-kaspersky-for-months-tapped-iran-nuke-talks/ Accessed on January 28, 2019.

4. Kaspersky Finds New Nation-State Attack – in Its Own Network. By Kim Zetter. Wired. June 10, 2015. https://www.wired.com/2015/06/kaspersky-finds-new-nation-state-attack-network/ Accessed on January 28, 2019.

5. Israel Spied on Iran Nuclear Talks with U.S. By Adam Entous. The Wall Street Journal. March 23, 2015. https://www.wsj.com/articles/israel-spied-on-iran-talks-1427164201 Accessed on January 27, 2019.

6. Spy Virus Linked to Israel Targeted Hotels Used for Iran Nuclear Talks. By Adam Entous and Danny Yadron. The Wall Street Journal. June 10, 2015. https://www.wsj.com/articles/spy-virus-linked-to-israel-targeted-hotels-used-for-iran-nuclear-talks-1433937601 Accessed on January 27, 2019.

7. Duqu 2.0: Reemergence of an aggressive cyberespionage threat. By Symantec Security Response. Symantec. June 10, 2015.

https://www.symantec.com/connect/blogs/duqu-20-reemergence-aggressive-cyberespionage-threat Accessed on January 27, 2019.

8. Шпионская история [A Spy Story]. Live Journal. By Eugene Kaspersky. Personal blog of Eugene Kaspersky. June 10, 2015. https://e-kaspersky.livejournal.com/234618.html Accessed May 4, 2019.

Chapter 2.

9. Countdown to Zero Day Countdown to Zero Day: Stuxnet and the Launch of the World's First Digital Weapon. By Kim Zetter. Crown Publishers. 2014. https://www.amazon.com/Countdown-Zero-Day-Stuxnet-Digital/dp/077043617X

10. Son of Stuxnet: The Digital Hunt for Duqu, a Dangerous and Cunning U.S.-Israeli Spy Virus. By Kim Zetter. The Intercept. November 12, 2014. https://theintercept.com/2014/11/12/stuxnet/ Accessed on January 27, 2019.

11. Duqu: Status Updates Including Installer with Zero-Day Exploit Found. By Vikram Thakur. Security Response. Symantec official blog. November 3, 2011. https://www.symantec.com/connect/w32-duqu_status-updates_installer-zero-day-exploit Accessed January 23, 2019.

12. Microsoft SMB Protocol and CIFS Protocol Overview. Microsoft Docs. Microsoft. May

30, 2018.
https://docs.microsoft.com/en-us/windows/win32/fileio/microsoft-smb-protocol-and-cifs-protocol-overview
Accessed January 23, 2019.

13. NetLock. Company profile.
https://onlinessl.netlock.hu/en/about-us/company-profile.html Accessed
March 1, 2019.

14. Belgium data center shuts down Duqu
server. By DataCenterDynamics. DDC.
November 04, 2011.
https://www.datacenterdynamics.com/news/belgium-data-center-shuts-down-duqu-server/ Accessed on January 29,
2019.

15. Duqu: father, son, or unholy ghost of
Stuxnet? By Jeremy Sparks, Robert M.
Lee, and Paul Brandau, cyberspace
officers. SC Magazine. SC Media.
November 2, 2011.
https://www.scmagazine.com/home/opinions/duqu-father-son-or-unholy-ghost-of-stuxnet/ Accessed August 23,
2019.

16. Kaspersky Lab Experts Discover
Unknown Programming Language in the
Duqu Trojan. Press release. Kaspersky
Lab. March 7, 2012.
https://usa.kaspersky.com/about/press-releases/2012_kaspersky-lab-experts-discover-unknown-programming-language-in-the-duqu-trojan Accessed
August 23, 2019.

Chapter 3.

17. The Duqu 2.0. Technical Details. Version 2.1. Kaspersky Lab. June 11, 2015. https://media.kasperskycontenthub.com/wp-content/uploads/sites/43/2018/03/07205202/The_Mystery_of_Duqu_2_0_a_sophisticated_cyberespionage_actor_returns.pdf Accessed on January 28, 2019.

18. Task Scheduler. Microsoft Docs. Microsoft. May 30, 2018. https://docs.microsoft.com/en-us/windows/win32/taskschd/task-scheduler-start-page Accessed on July 15, 2019.

19. User mode and kernel mode. Microsoft Docs. Microsoft. April 19, 2017. https://docs.microsoft.com/en-us/windows-hardware/drivers/gettingstarted/user-mode-and-kernel-mode Accessed on July 15, 2019.

20. API Index. Microsoft Docs. Microsoft. April 18, 2019. https://docs.microsoft.com/en-us/windows/win32/apiindex/api-index-portal Accessed on July 15, 2019.

21. Duqu 2.0: Reemergence of an aggressive cyberespionage threat. By Symantec Security Response. Symantec. June 10, 2015. https://www.symantec.com/connect/blogs/duqu-20-reemergence-aggressive-cyberespionage-threat Accessed on January 27, 2019.

22. Stepson of Stuxnet stalked Kaspersky for months, tapped Iran nuke talks. By Dan Goodin. Ars Technica. June 10, 2015. https://arstechnica.com/information-technology/2015/06/stepson-of-stuxnet-stalked-kaspersky-for-months-tapped-iran-nuke-talks/ Accessed on January 28, 2019.

Chapter 4.

23. Stepson of Stuxnet stalked Kaspersky for months, tapped Iran nuke talks. By Dan Goodin. Ars Technica. June 10, 2015. https://arstechnica.com/information-technology/2015/06/stepson-of-stuxnet-stalked-kaspersky-for-months-tapped-iran-nuke-talks/ Accessed on January 29, 2019.

24. US and British Spies Targeted Antivirus Companies. By Kim Zetter. WIRED. June 22, 2015. https://www.wired.com/2015/06/us-british-spies-targeted-antivirus-companies/ Accessed on January 29, 2019.

25. Popular Security Software Came Under Relentless NSA and GCHQ Attacks. By Andrew Fishman and Morgan Marquis-Boire. The Intercept. June 22, 2015. https://theintercept.com/2015/06/22/nsa-gchq-targeted-kaspersky/ Accessed on June 23, 2015.

26. GCHQ Application for Renewal of Warrant GPW/1160. By GCHQ. The Intercept. June 22, 2015. https://theintercept.com/document/2015/06/22/gchq-warrant-renewal/ Accessed on January 29, 2019.

Chapter 5.

27. Duqu and Destiny. By Donny Weber. Blogs. The Jerusalem Post. June 26, 2015. https://www.jpost.com/Blogs/Impact-in-Rehovot/Duqu-and-Destiny-407246 Accessed on April 14, 2019.

28. Top Israeli Security Officials Denies Responsibility for Cyber Attack Against Iran. JP Updates.com. June 25, 2015. http://jpupdates.com/2015/06/25/top-israeli-security-officials-denies-responsibility-for-cyber-attack-against-iran/ Accessed June 30, 2015.

29. Former defense officials doubt Israel behind Iran spying. By Yonah Jeremy Bob. The Jerusalem Post. June 25, 2015. https://www.jpost.com/Israel-News/Former-defense-officials-doubt-Israel-behind-Iran-spying-407073 Accessed April 14, 2019.

30. Russian Hackers Stole NSA Data on U.S. Cyber Defense. By Gordon Lubold and Shane Harris. The Wall Street Journal. October 5, 2017.

https://www.wsj.com/articles/russian-hackers-stole-nsa-data-on-u-s-cyber-defense-1507222108 Accessed April 20, 2019

31. How Israel Caught Russian Hackers Scouring the World for U.S. Secrets. By Nicole Perlroth and Scott Shane. The New York Times. Oct. 10, 2017. https://www.nytimes.com/2017/10/10/technology/kaspersky-lab-israel-russia-hacking.html Accessed April 7, 2019

32. Russia Has Turned Kaspersky Software Into Tool for Spying. By Shane Harris and Gordon Lubold. The Wall Street Journal. October 11, 2017. https://www.wsj.com/articles/russian-hackers-scanned-networks-world-wide-for-secret-u-s-data-1507743874 Accessed April 20, 2019.

33. Kaspersky Lab Antivirus Software Is Ordered Off U.S. Government Computers. By Matthew Rosenberg and Ron Nixon. The New York Times. Sept. 13, 2017. https://www.nytimes.com/2017/09/13/us/politics/kaspersky-lab-antivirus-federal-government.html?module=inline Accessed April 20, 2019.

34. Germany: 'No evidence' Kaspersky software used by Russians for hacks. By Andrea Shalal and Thorsten Severin. Reuters. October 11, 2017.

https://www.reuters.com/article/us-usa-security-kaspersky-germany/germany-no-evidence-kaspersky-software-used-by-russians-for-hacks-idUSKBN1CG284 Accessed April 20, 2019.

35. Israel's Kaspersky Hack Reveals That Russia Is Crossing All the Red Lines. By Anshel Pfeffer. HAARETZ. Oct 13, 2017. https://www.haaretz.com/israel-news/israel-s-kaspersky-hack-reveals-russia-crossing-all-red-lines-1.5457136 Accessed April 22, 2019.

Chapter 6.

36. Kaspersky Lab response to the alleged incident reported by the Wall Street Journal in an article published on October 5, 2017. Press release. Kaspersky Lab. October 5, 2017. https://usa.kaspersky.com/about/press-releases/2017_kaspersky-lab-response-to-the-alleged-incident-reported-by-the-wall-street-journal-in-an-article-published-on-october-5-2017 Accessed May 5, 2019.

37. Что это было? [What was that?]. Live Journal. By Eugene Kaspersky. Personal blog of Eugene Kaspersky. October 17, 2017. https://e-kaspersky.livejournal.com/440152.html Accessed May 4, 2019.

38. Russian anti-virus mogul: We uploaded US documents but quickly deleted them. By Raphael Satter. The Times of Israel. October 25, 2017. https://www.timesofisrael.com/russian-anti-virus-mogul-we-uploaded-us-documents-but-quickly-deleted-them/ Accessed June 10, 2019.

39. Investigation Report for the September 2014 Equation malware detection incident in the US. By Kaspersky Lab. November 16, 2017. https://securelist.com/investigation-report-for-the-september-2014-equation-malware-detection-incident-in-the-us/83210/tigation Report for the September 2014 Equation malware detection incident in the US Accessed April 22, 2019

40. The TeamSpy Crew Attacks – Abusing TeamViewer for Cyberespionage. By GReAT. Securelist.com. March 20, 2013. https://securelist.com/the-teamspy-crew-attacks-abusing-teamviewer-for-cyberespionage-8/35520/ Accessed May 11, 2019.

Chapter 7.

41. NSA employee who worked on hacking tools at home pleads guilty to spy charge. By Ellen Nakashima. The Washington Post. December 1, 2017.

https://www.washingtonpost.com/world/national-security/nsa-employee-who-worked-on-hacking-tools-at-home-pleads-guilty-to-spy-charge/2017/12/01/ec4d6738-d6d9-11e7-b62d-d9345ced896d_story.html?noredirect=on&utm_term=.7b710b5af1a2 Accessed May 11, 2019.

42. Here's the NSA Agent Who Inexplicably Exposed Critical Secrets. By Lily Hay Newman. Wired Magazine. December 1, 2017. https://www.wired.com/story/nsa-agent-exposed-critical-secrets/ Accessed April 24, 2019.

43. Former NSA software developer sentenced for keeping classified materials. By Andrew Blake. The Washington Times. September 25, 2018. https://www.washingtontimes.com/news/2018/sep/25/nghia-hoang-pho-former-nsa-developer-sentenced-ove/ Accessed April 27, 2019.

44. How Israel Caught Russian Hackers Scouring the World for U.S. Secrets. By Nicole Perlroth and Scott Shane. The New York Times. Oct. 10, 2017. https://www.nytimes.com/2017/10/10/technology/kaspersky-lab-israel-russia-hacking.html Accessed April 7, 2019

Chapter 8.

45. Russia's Top Cyber Sleuth Foils US Spies, Helps Kremlin Pals. By Noah Shachtman.

WIRED Magazine. July 23, 2012. https://www.wired.com/2012/07/ff_kaspersky/ Accessed March 8, 2019.

46. Who's Afraid of Kaspersky? By Lorenzo Franceschi-Bicchierai. MOTHERBOARD. May 22, 2018. https://motherboard.vice.com/en_us/article/wjbda5/kaspersky-sas-conference-russia-spying Accessed March 7, 2019.

47. The Company Securing Your Internet Has Close Ties to Russian Spies. By Carol Matlack, Michael Riley, and Jordan Robertson. Bloomberg News. March 19, 2015. https://www.bloomberg.com/news/articles/2015-03-19/cybersecurity-kaspersky-has-close-ties-to-russian-spies Accessed April 7, 2019.

48. A practical guide to making up a sensation. By Eugene Kaspersky. Nota Bene (Personal Blog). March 20, 2015. https://eugene.kaspersky.com/2015/03/20/a-practical-guide-to-making-up-a-sensation/ Accessed May 21, 2019.

49. Inside the Fight For The Soul Of Kaspersky Lab. By Ilya Zhegulev. BuzzFeedNews. January 22, 2018. https://www.buzzfeednews.com/article/ilyazhegulev/russia-kaspersky-antivirus Accessed May 18, 2019.

50. Kaspersky Lab: Based In Russia, Doing Cybersecurity In The West. By Corey Flintoff. National Public Radio (NPR). August 10, 2015.

https://www.npr.org/sections/alltechcon
sidered/2015/08/10/431247980/kasper
sky-lab-a-cybersecurity-leader-with-ties-
to-russian-govt Accessed May 18, 2019.

51. Why U.S. Officials Are Worried About This
Russian Cybersecurity Firm. By Jordan
Robertson and Aki Ito. Bloomberg News.
July 11, 2017.
https://www.bloomberg.com/news/articl
es/2017-07-11/why-u-s-officials-are-
worried-about-this-russian-cybersecurity-
firm Accessed on May 18, 2019.

52. Kaspersky Lab Has Been Working With
Russian Intelligence. By Jordan
Robertson and Michael Riley. Bloomberg
News. July 11, 2017.
https://www.bloomberg.com/news/articl
es/2017-07-11/kaspersky-lab-has-been-
working-with-russian-intelligence
Accessed on May 18, 2019.

53. Kaspersky Lab response clarifying the
inaccurate statements published in a
Bloomberg Businessweek article on July
11, 2017. Kaspersky Lab. July 11, 2017.
https://usa.kaspersky.com/about/press-
releases/2017_kaspersky-lab-response-
clarifying-inaccurate-statements-
published-in-bloomberg-businessweek-
on-july-11-2017 Accessed on May 18,
2019.

54. Honestly evaluating the Kaspersky
debate. By RenditionSec. Rendition
Infosec. July 12, 2017.

https://www.renditioninfosec.com/2017/07/honestly-evaluating-the-kaspersky-debate/ Accessed on May 18, 2019.

55. Exclusive: How a Russian firm helped catch an alleged NSA data thief. By Kim Zetter. POLITICO. 01/09/2019. https://www.politico.com/story/2019/01/09/russia-kaspersky-lab-nsa-cybersecurity-1089131 Accessed April 7, 2019.

56. The confrontation that fueled the fallout between Kaspersky and the U.S. government. Written by Patrick Howell O'Neill. CyberScoop. October 10, 2017. https://www.cyberscoop.com/kaspersky-fbi-cia-fsb-demarche-2015/ Accessed April 7, 2019.

57. Kaspersky in focus as US-Russia cyber-tensions rise. By Rob Lever. Times of Israel. 12 October 2017. https://www.timesofisrael.com/kaspersky-in-focus-as-us-russia-cyber-tensions-rise/ Accessed April 7, 2019.

Chapter 9.

58. Open Hearing on Worldwide Threats. Senate Hearing 115-205. U.S. Senate Select Committee on Intelligence. May 11, 2017. https://www.intelligence.senate.gov/hearings/open-hearing-worldwide-threats-hearing-0# Accessed on June 8, 2019.

59. I'm Eugene Kaspersky, cybersecurity guy and CEO of Kaspersky Lab! Ask me Anything! By Eugene Kaspersky. Reddit. May 11, 2017. https://www.reddit.com/r/IAmA/comments/6ajstf/im_eugene_kaspersky_cybersecurity_guy_and_ceo_, of/ Accessed on June 8, 2019.

60. Congress Casts A Suspicious Eye On Russia's Kaspersky Lab. By David Welna. NPR. July 5, 2017. https://www.npr.org/sections/parallels/2017/07/05/535651597/congress-casts-a-suspicious-eye-on-russias-kaspersky-lab Accessed on June 8, 2019.

61. U.S. senators seek military ban on Kaspersky Lab products amid FBI probe. By Dustin Volz, Joseph Menn. Reuters. June 28, 2017. https://www.reuters.com/article/us-kasperskylab-probe-idUSKBN19J2IX Accessed on June 8, 2019.

62. Kaspersky willing to turn over source code to US government. By Joe Uchill. The Hill. 07/02/2017. https://thehill.com/policy/cybersecurity/340420-kaspersky-willing-to-turn-over-source-code-to-us-government Accessed on June 1, 2019.

63. Why a Kaspersky code audit doesn't really ensure security. By RenditionSec.

ReinditionInfoSec Blog. July 5, 2017. https://blog.renditioninfosec.com/2017/07/why-a-kaspersky-code-audit-doesnt-really-ensure-security/ Accessed on June 1, 2019.

64. Russia Threatens Retaliation If US Bans Kaspersky Lab. By Mathew J. Schwartz. DataBreachToday. July 4, 2017. http://www.databreachtoday.com/russia-threatens-retaliation-if-us-bans-kaspersky-lab-a-10081 Accessed on June 1, 2019.

65. Twitter Post. By Robert M. Lee. Twitter. June 28, 2017. https://twitter.com/RobertMLee/status/880266103141826562 Accessed on June 8, 2019.

66. Russia Threatens Retaliation If Pentagon Bans Kaspersky Software. By Stepan Kravchenko. Bloomberg News. June 30, 2017. https://www.bloomberg.com/news/articles/2017-06-30/russia-threatens-retaliation-if-pentagon-bans-kaspersky-software-j4k2inwq Accessed on June 8, 2019.

Chapter 10.

67. Why U.S. Officials Are Worried About This Russian Firm. By Jordan Robertson and Aki Ito. Bloomberg News. July 11, 2017.

https://www.bloomberg.com/news/articles/2017-07-11/why-u-s-officials-are-worried-about-thi, s-russian-cybersecurity-firm Accessed May 18, 2019.

68. Kaspersky Lab Has Been Working With Russian Intelligence. By Jordan Robertson and Michael Riley. Bloomberg News. July 11, 2017. https://www.bloomberg.com/news/articles/2017-07-11/kaspersky-lab-has-been-working-with-russian-intelligence Accessed May 18, 2019.

69. What is a DDoS Attack? - DDoS Meaning. Kaspersky Lab. https://usa.kaspersky.com/resource-center/threats/ddos-attacks Accessed on June 1, 2019.

70. Kaspersky Lab response clarifying the inaccurate statements published in a Bloomberg Businessweek article on July 11, 2017. Kaspersky Lab. July 11, 2017. https://usa.kaspersky.com/about/press-releases/2017_kaspersky-lab-response-clarifying-inaccurate-statements-published-in-bloomberg-businessweek-on-july-11-2017 Accessed on May 18, 2019.

71. Honestly Evaluating the Kaspersky Debate. By RenditionSec. ReinditionInfoSec Blog. July 12, 2017. https://www.renditioninfosec.com/2017/07/honestly-evaluating-the-kaspersky-debate/ Accessed on June 1, 2019.

Chapter 11.

72. The confrontation that fueled the fallout between Kaspersky and the U.S. government. By Patrick Howell O'Neill. CyberScoop. October 10, 2017. https://www.cyberscoop.com/kaspersky-fbi-cia-fsb-demarche-2015/ Accessed on April 9, 2019.

Chapter 12.

73. Popular Security Software Came Under Relentless NSA and GCHQ Attacks. By Andrew Fishman, Morgan Marquis-Boire. The Intercept. 06/22/2015. https://theintercept.com/2015/06/22/nsa-gchq-targeted-kaspersky/ Accessed on June 23, 2015.

74. Equation Group: The Crown Creator of Cyber-Espionage. Press release. Kaspersky Lab. February 16, 2015. https://www.kaspersky.com/about/press-releases/2015_equation-group-the-crown-creator-of-cyber-espionage Accessed on June 14, 2019.

75. *Equation Group: Questions and Answers (Version: 1.5)*. Kaspersky Lab. SecureList.com. February 2015. https://securelist.com/files/2015/02/Equation_group_questions_and_answers.pdf Accessed on June 14, 2019.

76. Exclusive: How a Russian firm helped catch an alleged NSA data thief. By Kim Zetter. POLITICO. 01/09/2019.

https://www.politico.com/story/2019/01/09/russia-kaspersky-lab-nsa-cybersecurity-1089131 Accessed April 7, 2019.

77. Twitter post. By Edward Snowden. Twitter. August 16, 2016. https://twitter.com/Snowden/status/765514891813945344 Accessed April 7, 2019.

78. Kaspersky Lab uncovers Slingshot, the spy that came in from the router. Press release. Kaspersky Lab. March 9, 2018. https://usa.kaspersky.com/about/press-releases/2018_slingshot Accessed June 23, 2019.

79. The Slingshot APT FAQ. By Alexey Shulmin, Sergey Yunakovsky, Vasily Berdnikov, Andrey Dolgushev. SecureList.com. March 9, 2018. https://securelist.com/apt-slingshot/84312/ Accessed June 23, 2019.

80. Who's Afraid of Kaspersky? By Lorenzo Franceschi-Bicchierai. MOTHERBOARD. May 22, 2018. https://motherboard.vice.com/en_us/article/wjbda5/kaspersky-sas-conference-russia-spying Accessed March 7, 2019.

81. Twitter post. By Michael Rea. Twitter. March 21, 20018. https://twitter.com/ComradeCookie/status/976633797419581446 Accessed March 7, 2019.

82. Kaspersky in focus as US-Russia cyber-tensions rise._ By Rob Lever. Times of

Israel. 12 October 2017.
https://www.timesofisrael.com/kaspersk
y-in-focus-as-us-russia-cyber-tensions-
rise/ Accessed April 7, 2019.

83. Kaspersky: Yes, we obtained NSA secrets.
No, we didn't help steal them. By Dan
Goodin. Ars Technica. November 16,
2017.
https://arstechnica.com/information-
technology/2017/11/kaspersky-yes-we-
obtained-nsa-secrets-no-we-didnt-help-
steal-them/ Accessed on May 26, 2019.

Chapter 13.

84. Trust First: Kaspersky Lab launches its
Global Transparency Initiative; will
provide source code – including updates –
for a third-party review; will open three
Transparency Centers worldwide. Press
release. Kaspersky Lab. October 23,
2017.
https://usa.kaspersky.com/about/press-
releases/2017_trust-first-kaspersky-lab-
launches-its-global-transparency-
initiative Accessed June 1, 2019.

85. Kaspersky relocates data processing to
Switzerland. Kaspersky Lab. Official website: www.kaspersky.com.
https://www.kaspersky.com/transparenc
y-center. Accessed July 8, 2019.

86. Trump signs into law U.S. government
ban on Kaspersky Lab software. By
Dustin Volz. Reuters. December 12, 2017.

https://www.reuters.com/article/us-usa-cyber-kaspersky/trump-signs-into-law-u-s-government-ban-on-kaspersky-lab-software-idUSKBN1E62V4?il=0 Accessed July 8, 2019.

87. An Open Letter from Kaspersky Lab. By Kaspersky Lab. Kaspersky Lab. December 18, 2017. https://www.kaspersky.com/blog/kaspersky-lab-open-letter/20501/ Accessed June 27, 2019.

88. Russia-Based Kaspersky Lab Sues Trump Administration For Banning Its Software. By Emily Sullivan. NPR. December 18, 2017. https://www.npr.org/sections/thetwo-way/2017/12/18/571710070/russia-based-kaspersky-lab-sues-trump-administration-for-banning-its-software Accessed June 27, 2019.

89. Kaspersky Lab Files Another Lawsuit in Wake of NDAA Ban. By Chris Bing. CYBERSCOOP. February 13, 2018. https://www.cyberscoop.com/kaspersky-lawsuit-ndaa-ban/ Accessed July 8, 2019.

90. Kaspersky Lab lawsuits against US thrown out. By Abrar Al-Heeti. CNET. May 30, 2018. https://www.cnet.com/news/kaspersky-lab-lawsuits-against-us-thrown-out/ Accessed June 27, 2019.

91. Memorandum Opinion: Kaspersky Lab, Inc. et al v. United State of America. Opinion from US District Judge for the

District of Columbia Colleen Kollar-Kotelly. Uploaded by Jonathan Skillings. SCRIBD.com. May 30, 2018. https://www.scribd.com/document/3805 95629/MEMORANDUM-OPINION-KASPERSKY-LAB-INC-et-al-v-UNITED-STATES-OF-AMERICA#from_embed Accessed July 8, 2019.

Chapter 14.

92. The confrontation that fueled the fallout between Kaspersky and the U.S. government. By Patrick Howell O'Neill. CyberScoop. October 10, 2017. https://www.cyberscoop.com/kaspersky-fbi-cia-fsb-demarche-2015/ Accessed on April 9, 2019.

93. How Israel Caught Russian Hackers Scouring the World for U.S. Secrets. By Nicole Perlroth and Scott Shane. The New York Times. Oct. 10, 2017. https://www.nytimes.com/2017/10/10/t echnology/kaspersky-lab-israel-russia-hacking.html Accessed April 7, 2019

94. Russia Has Turned Kaspersky Software Into Tool for Spying. By Shane Harris and Gordon Lubold. The Wall Street Journal. October 11, 2017. https://www.wsj.com/articles/russian-hackers-scanned-networks-world-wide-for-secret-u-s-data-1507743874 Accessed April 20, 2019.

95. Memorandum Opinion: Kaspersky Lab, Inc. et al v. United State of America. Opinion from US District Judge for the District of Columbia Colleen Kollar-Kotelly. Uploaded by Jonathan Skillings. SCRIBD.com. May 30, 2018. https://www.scribd.com/document/380595629/MEMORANDUM-OPINION-KASPERSKY-LAB-INC-et-al-v-UNITED-STATES-OF-AMERICA#from_embed Accessed July 8, 2019.

96. Kaspersky Lab announces 4% revenue growth to $726 million in 2018. Press release. Kaspersky Lab. February 19, 2019. https://www.kaspersky.com/about/press-releases/2019_kaspersky-lab-announces-4-percent-revenue-growth-to-726-million-dollars-in-2018 Accessed August 22, 2019.

97. Что это было? [What was that?]. Live Journal. By Eugene Kaspersky. Personal blog of Eugene Kaspersky. October 17, 2017. https://e-kaspersky.livejournal.com/440152.html Accessed May 4, 2019.

98. Who's Afraid of Kaspersky? By Lorenzo Franceschi-Bicchierai. MOTHERBOARD. May 22, 2018. https://motherboard.vice.com/en_us/article/wjbda5/kaspersky-sas-conference-russia-spying Accessed March 7, 2019.

99. Undercover spy targeted Kaspersky Lab critics: Report. By Raphael Satter.

Associated Press. April 17, 2019. Washington Times. https://www.washingtontimes.com/news/2019/apr/17/ap-exclusive-undercover-spy-targeted-kaspersky-cri/ Accessed June 10, 2019.

100. AP Exclusive: Mysterious operative haunted Kaspersky critics. By Raphael Satter. Associated Press. April 17, 2019. Washington Times. https://www.washingtontimes.com/news/2019/apr/17/ap-exclusive-private-spy-targeted-critics-of-kaspe/ Accessed June 10, 2019.

101. Israeli firm Black Cube denies link to spy targeting Kaspersky Lab critics. By Raphael Satter. Associated Press. April 17, 2019. The Times of Israel. https://www.timesofisrael.com/israeli-firm-black-cube-denies-link-to-spy-targeting-kaspersky-lab-critics/ Accessed June 10, 2019.

102. Is your antivirus software part of your threat model? Maybe it should be… By RenditionSec. ReinditionInfoSec.com. July 11, 2017. https://blog.renditioninfosec.com/2017/07/av_threat_model_kaspersky/ Accessed June 1, 2019.

103. Should Antivirus software be part of your threat model? By RenditionSec. ReinditionInfoSec.com. October 8, 2017. https://www.renditioninfosec.com/2017/10/should-antivirus-software-be-part-of-your-threat-model/ Accessed June 1,

2019.

104. Bolstering the Government's Cybersecurity: Assessing the Risk of Kaspersky Lab Products to the Federal Government. Hearing before the U.S. House of Representatives Subcommittee on Oversight and Committee on Science, Space, and Technology. Serial No. 115–33. October 25, 2017. https://docs.house.gov/meetings/SY/SY21/20171025/106556/HHRG-115-SY21-20171025-SD003.pdf Accessed on July 10, 2019.

105. Would you trust Eugene Kaspersky, Russia's 'Cyber Security King'? By Claire Williams, Farah Boucherak. France24. June 10, 2015. http://www.france24.com/en/20151006-eugene-kaspersky-trust-russia-cyber-security-king-spy-espionage-kgb-allegations-banya-monac Accessed on October 7, 2015.

106. Kaspersky relocates data processing to Switzerland. Kaspersky Lab. Official website: www.kaspersky.com. https://www.kaspersky.com/transparency-center. Accessed July 8, 2019.

107. Kaspersky Lab Opens New Transparency Center in Madrid and Conducts Independent Legal Assessment of Russian Legislation Related to Data Processing. Kaspersky Lab. Press release. April 2, 2019.

https://usa.kaspersky.com/about/press-releases/2019_transparency-center-in-madrid Accessed on July 10, 2019.

108. Kaspersky touts APAC Transparency Center as proving 100% trustworthiness. By Asha Barbaschow. ZD Net. August 16, 2019. https://www.zdnet.com/article/kaspersky-touts-apac-transparency-center-as-proving-100-trustworthiness/ Accessed on August 22, 2019.

109. Building a safer world with Kaspersky: The company unveils new branding and visual identity. Press release. Kaspersky. June 4, 2019. https://www.kaspersky.com/about/press-releases/2019_the-company-unveils-new-branding-and-visual-identity Accessed July 12, 2019.

110. Kasper-Spy: Kaspersky Anti-Virus puts users at risk. By Ronald Eikenberg. c't Magazine. August 15, 2019. https://www.heise.de/ct/artikel/Kasper-Spy-Kaspersky-Anti-Virus-puts-users-at-risk-4496138.html Accessed on August 22. Accessed on August 22, 2019.

111. Vulnerability Report: List of Advisories. Advisory issued on 11th July 2019. Kaspersky Lab. July 11, 2019. https://support.kaspersky.com/general/vulnerability.aspx?el=12430#110719 Accessed on August 22, 2019.

112. Duqu 1.5: A Ghost in the Wires of a Diplomatic Venue. By J. A. Guerrero-Saade and Silas Cutler. Chronicle. April 9, 2019. https://storage.googleapis.com/chronicle-research/DuQu%201.5%20A%20Ghost%20in%20the%20Wires.pdf Accessed April 9, 2019.